Field Manual
No. 5-10

Headquarters
Department of the Army
Washington, DC, 3 October 1995

# Combat Engineer

## TABLE OF CONTENTS



---

Page

# PREFACE

*Field Manual (FM) 5-10* is designed as a companion manual for *FM 5-34* and is structured using the following Battlefield Operating Systems (BOS) as a framework:

- Intelligence.
- Maneuver.
- Mobility and survivability.
- Fire support (FS).
- Air defense (AD).
- Combat service support (CSS).
- Command and control ($C^2$).

Each BOS is not discussed in detail. Instead, they are used to outline considerations for how the engineer platoon supports the total combined-arms (CA) effort. The bulk of the platoon's missions concentrates on mobility, countermobility, and survivability operations in support of the mobility and survivability BOS. However, the platoon leader and sergeant must have an understanding of the role their platoon plays in supporting all the BOS. More importantly, the platoon leader and sergeant must understand the impact of all the BOS on platoon planning, preparation, and execution.

This FM also has three appendices: *Appendix A* covers basic formations, movement techniques, and hand-and-arm signals; *Appendix B* contains checklists for security, precombat, and packing; and *Appendix C* covers case studies for mounted and dismounted breaching.

---

The proponent of this publication is Headquarters (HQ), United States (US) Army Engineer School. Send comments and recommendations on *Department of the Army (DA) Form 2028* directly to Commander, US Army Engineer School, ATTN: ATSE-T-PD-PM, Fort Leonard Wood, Missouri 65473-6650.

Unless this publication states otherwise, masculine nouns and pronouns do not refer exclusively to men.

# CHAPTER 1
# Intelligence

Intelligence is the key for preparing for the next battle. The analysis of intelligence, in detail, determines the impact of the enemy, weather, and terrain on operations. It is a continuous process that supports the planning for and execution of all operations. It is imperative that the platoon leader understand the intelligence process and his part in that process.

## THREAT

The threat facing the US Army has gone through a quantum change with the breakup of the Warsaw Pact and the former Soviet Union. However, the majority of the forces we will face in mid- to high-intensity conflicts will use the "Soviet" model for their operations. *Figures 1-1 and 1-2, pages 1-2 and 1-3,* provide templates for the Soviet-style platoon and company defense as well as movement formations. A detailed analysis of threat mine-warfare capabilities and tactics can be found in *FM 90-13-1, Appendix A, and FM 20-32, Appendix C.*

## RECONNAISSANCE

Reconnaissance is critical to the engineer. During offensive operations, the engineer must see the battlefield to plan breaching operations. History has proven that when a unit conducts CA rehearsals and obstacle reconnaissance it has greater success. The platoon leader should organize his platoon based on the reconnaissance mission the commander gives him. This frequently means providing expert engineers to assist the task force (TF) scouts. The platoon leader may also organize all or part of his platoon for a reconnaissance mission if there is a critical requirement for engineer-specific intelligence. If at all possible, the engineers who conduct the reconnaissance should be trained in this task as well as cross trained with the maneuver unit's scout platoon. The platoon leader must ensure that his designated reconnaissance element links up with the scout platoon.

### *TYPES*

The general types of engineer reconnaissance missions are—

- Technical, which—
  - Is conducted to collect engineer-specific technical data.

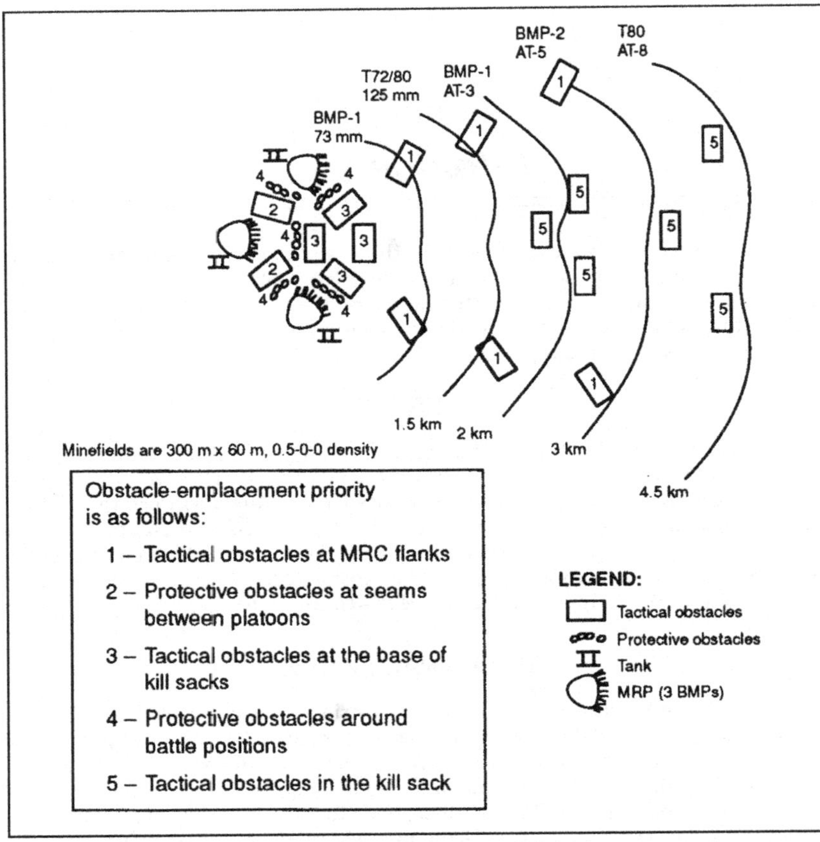


**Figure 1-1. Motorized rifle company (MRC) defense template**

  - Is conducted to perform a specified task from higher HQ in a Level 1 threat environment.
  - Involves (almost always) engineers only and is usually conducted behind the forward line of own troops/forward edge of the battle area (FLOT/FEBA).
- Tactical, which—
  - Is conducted to collect engineer-specific information requirements on an enemy point, area target, or attack axis or a route.

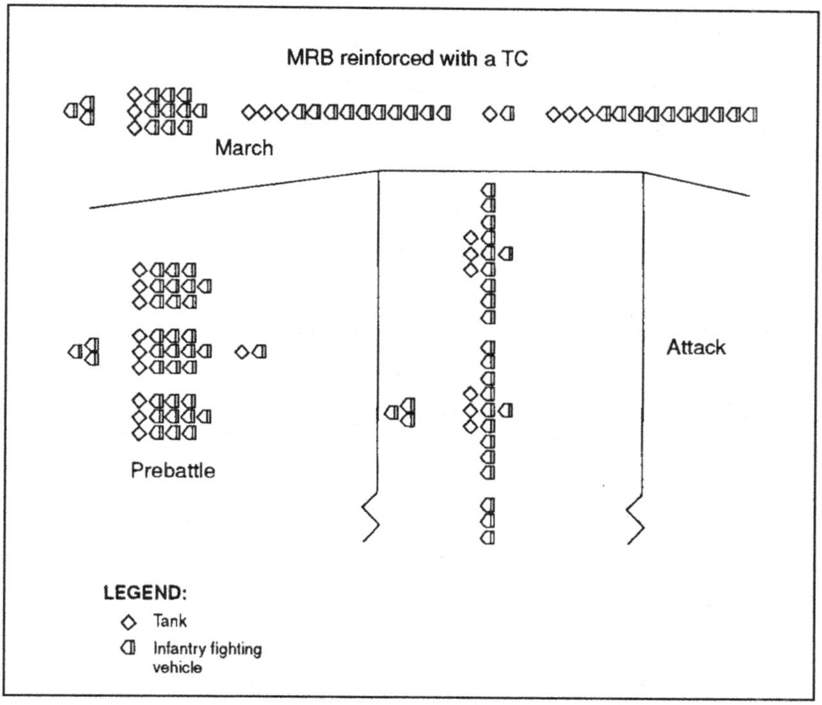


**Figure 1-2. Typical threat formations**

- Is conducted to support the maneuver plan as an implied task.
- Is conducted in a Level 3 threat environment.
- Requires engineer participation as part of a CA effort.
- Is usually conducted forward of the FLOT/FEBA.

### *PLANNING*

Reconnaissance is a CA effort. The engineers either augment the existing reconnaissance element or conduct their own reconnaissance as part of the consolidated reconnaissance and surveillance (R&S) plan. Engineer reconnaissance must be focused on a specific target, and the reconnaissance element should be given specific intelligence requirements.

In heavy forces, the engineer company commander provides the reconnaissance target and does the majority of the engineer reconnaissance planning. However, in light forces, the engineer platoon leader does the engineer reconnaissance planning.

Platoon reconnaissance planning should include—

- A warning order (WO) based on brigade and TF planning, which includes—
  - Tentative target locations.
  - Type of reconnaissance or target (see *Figure 1-3*).
  - Specific information, equipment, and coordination requirements.
- The movement of reconnaissance assets to effect the linkup.
- The maximum time for leaders to prepare for the reconnaissance mission.
- The assignment of detailed reconnaissance routes.
- The actions to be taken on contact.

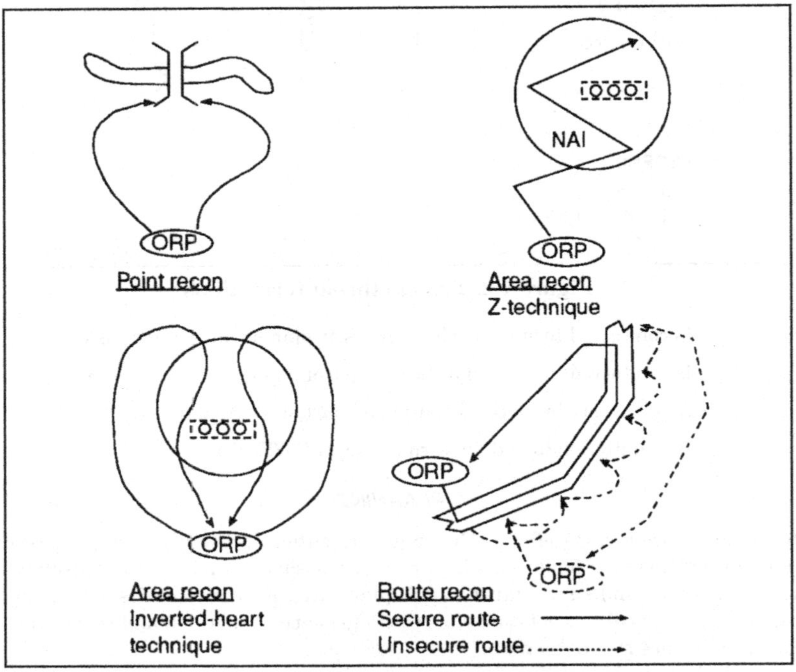


**Figure 1-3. Recon techniques**

- Consolidation instructions.
- Specific reporting requirements.

## MOUNTED AND DISMOUNTED PATROLLING

Patrolling is a CA operation. Engineers are normally part of a larger unit's patrol. When patrolling mounted—

- Position all reconnaissance assets forward.
- Use aggressive and continuous patrols.
- Use area and zone reconnaissance.
- Allow enough time for a thorough reconnaissance.
- Use stealth, not combat power, for reconnaissance.
- Modify formations to overwatch the point element.
- Use terrain driving techniques; use draws and folds in the terrain to mask movement.
- Use night, fog, dust, and storms to conceal movement.
- Reduce speeds to avoid dust signatures.
- Travel at the highest tactical speeds, and place artillery fires on expected enemy positions to avoid long-range fires.
- Keep vehicles dispersed.

When patrolling dismounted—

- Reconnoiter all danger areas, choke points, and critical terrain features.
- Expect contact with enemy dismounted scouts.
- Use darkness, storms, and periods of limited visibility to infiltrate enemy positions.
- Use binoculars or night-vision devices at night with the illumination from ambient light (ensure that laser protection is maintained).
- Use artillery illumination and spotting rounds on reconnaissance targets to assist in land navigation and orientation.

# CHAPTER 2
# Maneuver

Engineer support to maneuver units requires detailed planning. Engineers may be part of the maneuver unit's formations and, if so, will provide the combat multiplier of engineer effort. *FM 90-13-1* is the primary doctrinal manual for breach planning. *FMs 20-32 and 90-7* are the primary doctrinal manuals for obstacle planning and construction. These references provide the technical information required to plan engineer support for offensive and defensive maneuvers.

The platoon leader should have an understanding of the tactical decision-making process, troop-leading procedures (TLP), and orders preparation. *Figure 2-1, page 2-2,* shows the relationship between TLP, the estimate process, and the intelligence preparation of the battlefield (IPB) process.

In addition, the platoon leader must advise the maneuver commander on the military aspects of the terrain since he is the terrain expert. Normally, the platoon leader only has a map to prepare his terrain analysis. However, he should request terrain-analysis products from his parent engineer company. In both light and armored units, the engineer company has access to terrain-analysis and terrain-visualization products or the ability to develop them. These products greatly aid platoon-level planning.

## OFFENSE

The platoon leader is the key engineer expert to his supported unit. He must make several key decisions during the offense. He recommends where critical breach assets are located and which ones are best based on the situation. For example, the supporting engineer platoon leader, in an armored unit, would recommend where and how the M1 tank's plow would be used; or a light platoon leader could recommend to the TF commander a covert breach based on his terrain analysis. In both cases, the engineer platoon leader assists the maneuver commander with his planning. For more information on breaching, see *FM 90-13-1*. The following considerations are specific to the offense:

- Using a highly mobile engineer force, well forward and integrated into maneuver formations, to maintain the momentum of the attack.

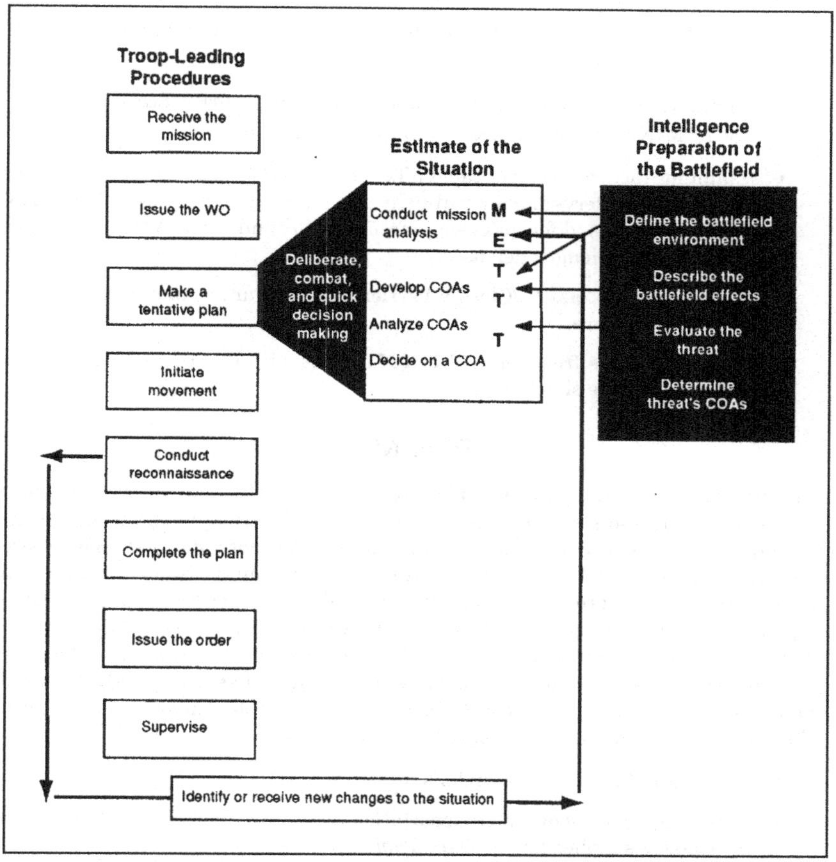


**Figure 2-1. Relationship between TLP, the estimate process, and the IPB process**

- Arranging for lane turnover between the forward breaching force and follow-on engineers for lane improvement and obstacle clearance.
- Ensuring that marking material and the standing operating procedure (SOP) for lane marking are coordinated between passing units.
- Ensuring that critical engineer equipment, such as armored vehicle-launched bridges (AVLBs), follow-on tactical bridging, lift capability for

line-charge reloading, and lane-marking materials to replenish marking systems, is available.

- Planning situational-obstacle capability for flank security. Scatterable mine (SCATMINE) systems work well in this application.
- Planning tactical-obstacle capability for protection against counterattacks (CATKs). Artillery- and aircraft-delivered family of scatterable mines (FASCAM) are planned for quick emplacement. Ground FASCAM systems are planned for prolonged defenses.
- Ensuring that Class IV/Class V barrier materials are available for a transition to the defense.
- Removing curbs from the AVLB if M1 tanks with rollers are to cross. The roller's dog bone should be raised.

## DEFENSE

As with the offense, the platoon leader recommends to the maneuver commander where engineer assets are best used. It is critical that the platoon leader be an integral part of obstacle planning and integration. For more information on obstacle integration, see *FM 90-7*. The platoon leader should also coordinate and develop the obstacle-protection plan. It is critical that the obstacles are protected against enemy breaching attempts, especially covert attempts. The platoon leader should coordinate the inspection of the obstacles that his platoon built before the expected enemy attack. This will ensure that the enemy has not covertly breached and, if so, allows for the obstacles to be repaired before the enemy's main attack. The following considerations are specific to the defense:

- Understanding the commander's intent for obstacles.
- Ensuring that adequate transportation assets are available to haul Class IV/Class V supplies forward (see *Table 2-1, page 2-4*).
- Identifying critical engineer tasks early. Terrain preparation requires time for completion. Engineers must not remain idle while planning is in progress. If possible, mortars, air-defense artillery (ADA), fire-support-team (FIST) vehicles, CSS assets, and the tactical operations center (TOC) should be dug in while direct-fire systems are being positioned. Typically, these systems can be fortified between issuing the WO and completing the direct-fire systems' positioning.
- Practicing operations security (OPSEC) measures to prevent premature disclosure of the defense.

**Table 2-1. Class IV/Class V haul capacities**

| Vehicle | Concertina Wire | M15 AT Mine | M21 AT Mine | M16 AP Mine | M14 AP Mine | MOPMS Mine | Flipper Mine | Volcano Mine | MICLIC Reload |
|---|---|---|---|---|---|---|---|---|---|
| HMMWV, M998<br>2,500 lb<br>215 cu ft | 2 | 51 | 27 | 55 | 56 | 15 | 11 | 1 | *1 |
| 2 1/2 -ton truck<br>5,000 lb<br>443 cu ft | 4 | 102 | 55 | 111 | 113 | 30 | 23 | 2 | 1 |
| 5-ton truck<br>10,000 lb<br>488 cu ft | 7 | 204 | 109 | 222 | 227 | 61 | 46 | 5 | 3 |
| 5-ton dump truck<br>10,000 lb<br>***135/291 cu ft | 2/4 | 112/204 | 32/70 | 168/222 | 71/153 | 23/51 | 39/46 | 3/5 | 2/3 |
| 20-ton dump truck 40,000 lb<br>754 cu ft | 11 | 628 | 179 | 888 | 443 | 132 | 184 | 20 | 11 |
| HEMTT truck<br>20,000 lb<br>540 cu ft | 8 | 408 | 128 | 444 | 317 | 94 | 92 | 10 | 7 |
| 12-ton S&P<br>24,000 lb<br>875 cu ft | 13 | 489 | 208 | 533 | 514 | 148 | 110 | 12 | 9 |
| 40-ton lowboy<br>80,000 lb<br>1,760 cu ft | 27 | 1,466 | 419 | 1,777 | 1,035 | 308 | 368 | 43 | 27 |
| M54812,000 lb<br>529 cu ft | 8 | 244 | 125 | 266 | 272 | 74 | 55 | 6 | 4 |
| #Mines / wt/lb<br>Cube cu ft | 40/ 1,180<br>64 | 1/ 49<br>1.2 | 4/ 91<br>4.1 | 4/ 45<br>.8 | 90/ 44<br>1.9 | 21/ 162<br>5.7 | 40/ 217<br>3.4 | 240/ 1,850<br>37.6 | ** 2,656<br>64.8 |

# for concertina = bundles; 1 bundle = 40 rolls   * Overloads vehicle   **Line charge + rocket   ***Without/with sideboards

- Ensuring that the engineer organization for combat allows for rapid transition to the offense. The reserve must always have a designated force of engineers. Obstacle placement must not interfere with spoiling and CATKs.
- Ensuring that engineer units are not held in reserve but remain committed and work on the commander's priority tasks.

## PLATOON LEADER'S PLANNING

The platoon leader can use *Tables 2-2 and 2-3, pages 2-6 and 2-7,* as a guideline when planning and conducting offensive and defensive operations.

## REHEARSAL PRINCIPLES

Regardless of the event or task to be rehearsed or the type of rehearsal used, certain principles are universal for conducting effective rehearsals. They are—

- Supporting the scheme of maneuver and the commander's intent.
- Providing clear tasks/conditions/standards (T/C/S).
- Conducting multiechelon CA rehearsals.
- Determining key participants.
- Enforcing standards/training to standard.
- Using the after-action-review (AAR) process to provide feedback to all participants.
- Complementing the preparation phase.
- Instilling confidence in the plan and in the leaders.

During the command estimate process, use the rehearsal principles. These principles produce the—

- Participant levels.
- Rehearsal technique.
- Initiation of precombat checks (PCCs), precombat inspections (PCIs), and mission-specific drills.

Proper application of the rehearsal principles enables the unit to progress through a "crawl/walk/run" process that ends with mission success (see *Figure 2-2, page 2-8*).

## Table 2-2. TF-engineer offense briefing

| Briefing Topic | Main Points to Brief |
|---|---|
| Engineer task organization | • How the platoon will be task-organized<br>• Special equipment that will be task-organized to subordinate units<br>• Times and locations for linkup |
| Enemy situation | • Obstacle template<br>• Enemy's obstacle intent<br>• Enemy's expected level of defensive preparation<br>• Confirmed intelligence on enemy obstacles and positions (include an analysis of this intelligence)<br>• Enemy employment of FASCAM<br>• Impact of enemy obstacles on friendly maneuver |
| OBSTINTEL collection effort | • Engineer R&S plan<br>• NAIs critical to OBSTINTEL<br>• How the OBSTINTEL effort supports the scheme of maneuver<br>• Scout, engineer, and maneuver OBSTINTEL responsibilities and relationships |
| Scheme of engineer operations | • How the platoon will support the scheme of maneuver from the objective to the LD, to include consolidation<br>• Engineer contingency plans and pertinent decision points |
| TF and company/team actions at obstacles | • In-stride breaching, to include—<br>– Areas where the unit can expect to in-stride breach<br>– Special equipment required<br>– Special instructions<br>• Deliberate breaching, to include—<br>– Transitioning from attack formations to breaching formations<br>– How the fundamentals of breaching will be synchronized<br>– Key events and control measures synchronizing the breach<br>– Composition of and instructions for the support, breach, and assault forces<br>– Scheme of fires supporting the breach<br>– Obscuration plan<br>– Coordinating instructions, command and signals for the breach<br>• Assault breaching, to include—<br>– Allocation of engineers<br>– Special coordinating instructions |
| Countermobility support to offensive maneuver | • Planned locations for situational obstacles<br>• Triggers for situational-obstacle employment |
| Coordinating instructions | • Lane marking, breach traffic control, order of march through the lane, specific breaching signals, and breach $C^2$ |
| Service support | • What support the platoon is receiving from the supported company/team and the parent company, if different from the SOP<br>• Medical evacuation, supplies, and vehicle recovery<br>• Current equipment status if critical to the mission |
| Unresolved issues | • Issues that the plan does not adequately address |

## Table 2-3. TF-engineer defense briefing

| Briefing Topic | Main Points to Brief |
|---|---|
| Engineer task organization | • How the platoon will be task-organized<br>• Special equipment that will be task-organized to subordinate units<br>• Times and locations for linkup |
| Enemy | • Enemy maneuver formations through the sector, to include—<br>– Most probable enemy COA<br>– Place and time the enemy will change formation<br>– Enemy breaching assets and capabilities (include high-priority targets) |
| Countermobility support | • Scheme of obstacles, to include—<br>– Obstacle zones and restrictions<br>– Obstacle belts and groups and their effects<br>– Desired effect of obstacle groups on enemy maneuver<br>– How the scheme of obstacles support the direct- and indirect-fire plans<br>– Triggers and decision points for situational and reserve obstacles<br>– Location of Class IV/ClassV supply points<br>• Obstacle status reporting, to include—<br>– Completion of obstacles<br>– Barrier-material status<br>• Obstacle-protection and turnover plans |
| Mobility support | • Where the lanes need to be to support MSRs, company/team supply routes, and reconnaissance passages<br>• Responsibility for lane siting and closure<br>• Lane marking and contact points<br>• Obstacle restrictions that support the TF's mobility<br>• Engineer-support plan for mobility |
| Survivability support | • Protective-obstacle plan, to include—<br>– Material that the unit allocates<br>– Any engineer assistance required<br>– Protective-obstacle reporting responsibilities<br>• Construction plan for fighting positions, to include—<br>– Any engineer assistance required to construct individual positions<br>– Allocation of engineer equipment and time and the priority for constructing vehicle positions<br>– Linkup points and times with the company/team<br>• Equipment control plan to ensure that maneuver companies know—<br>– How many positions will be constructed by type<br>– When, where, and how much equipment they will receive<br>– How long they will retain the equipment |
| Engineer time line | • Engineer time line, to include—<br>– Estimated completion time<br>– Enemy trigger to stop work<br>– Withdrawal route or passage<br>– Engineer consolidation plan<br>– Class VI/V supply-point recover or destruction |
| Coordinating instructions | • Obstacle-siting responsibilities<br>• Reporting times<br>• Class IV/Class V supply-point location, security plan, and responsibility |

Table 2-3. TF-engineer defense briefing (continued)

| Briefing Topic | Main Points to Brief |
|---|---|
| Service support | • Critical equipment and material shortages affecting defensive preparations<br>• Medical evacuation plan for engineers<br>• Barrier-material configurations to support obstacle-group construction |
| Unresolved issues | • Issues that the plan does not adequately address |

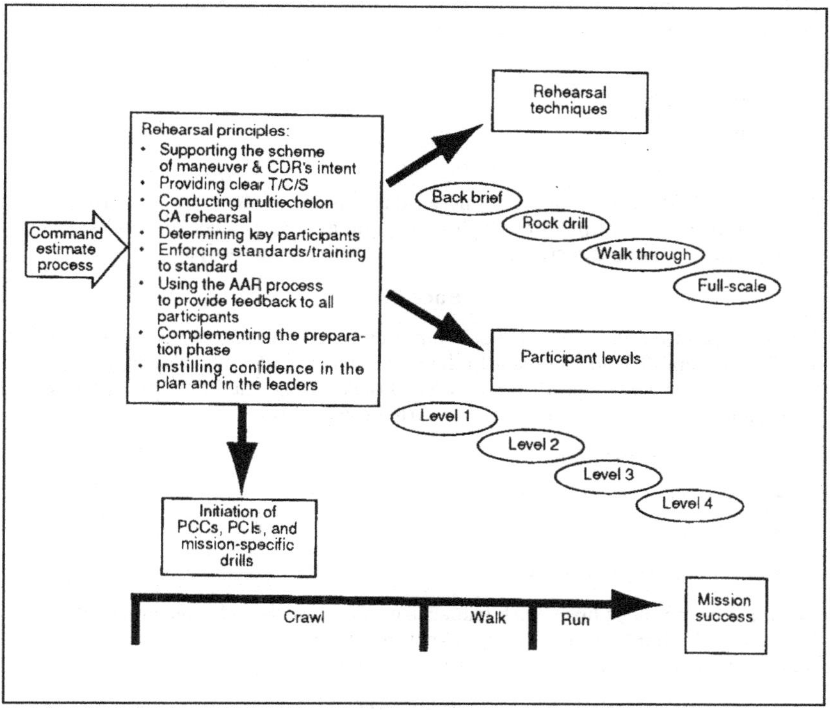


Figure 2-2. Rehearsal planning

### *PARTICIPANT LEVELS*

The participant level details exactly who in the unit is required to attend the rehearsal. A CA-team rehearsal may have four distinct levels. The platoon leader may participate with or without his subordinates in the CA rehearsal. The three levels that apply to a platoon rehearsal follow:

- Level 1 - Squad leaders and platoon HQ.
- Level 2 - Squad leader, team leader, special teams, and vehicle commanders.
- Level 3 - Entire unit.

### *REHEARSAL TECHNIQUES*

Rehearsal techniques fall into four different categories that follow the "crawl/walk/run" training concept. These categories are—

- Back brief.
- Rock drill.
- Walk through.
- Full-scale.

Each type of rehearsal reflects an increase in mission realism and a corresponding increase in rehearsal payoff. Each technique also increases the realism of the enemy, terrain, subunit actions, and time/distance relationships.

#### Back Brief

The back-brief rehearsal is an event that occurs after an operation order (OPORD) brief. Subordinate leaders repeat to the commander what he expects them to do and why. They identify all specified and implied tasks, determine their mission-essential tasks, and give their restated mission. During the back brief, they address the—

- Commander's intent.
- Concept of the operation.
- Scheme of maneuver.
- Timing to complete tasks.

During a typical back brief, each leader uses a map or a sand table and explains his mission. The back brief is the quickest of all the rehearsal techniques. It is a leader's tool and is typically the first rehearsal for the unit.

#### Rock Drill

A rock-drill rehearsal is the acting out of friendly and enemy actions based on the scheme of maneuver and the situation and event templates. Participants rehearse their actions by moving something that represents themselves or their unit, such as rocks or sticks. While acting out the plan, participants talk through their missions, critical tasks, actions, and decisions. Since all participants are simultaneously acting out their part of the scheme of maneuver, leaders can identify problems and disconnects in synchronization more clearly in a rock drill.

### Walk Through

A walk-through rehearsal is the acting out of the scheme of maneuver using mounted or dismounted movement based on the assets employed during the attack. Participants should also communicate with the same type of equipment they will use during the action. Participants rehearse by—

- Maneuvering their vehicles (mounted movement) or themselves (dismounted movement).
- Reporting critical actions.
- Making required decisions.

Since the participants are in a more realistic environment, they rehearse the finer aspects of synchronization, $C^2$, and subunit actions. Aggressive portrayals of enemy actions and reactions are critical in walk-through rehearsals. This threat portrayal is the backbone of increasing the rehearsal realism. This rehearsal technique is more difficult to orchestrate than the back brief or rock drill; however, it is the optimum balance between resource constraints and realism. Successfully mastering a walk-through rehearsal should be the minimum goal for all units.

### Full-Scale

During a full-scale rehearsal, participants use real-time mounted and dismounted movement over the actual or similar terrain. Typically, it is conducted at Level 3; however, a mix of participant levels is possible. At least one subunit must participate at Level 3 for a full-scale rehearsal. This type of rehearsal is obviously the most resource intensive, but it provides the most realistic training environment for the unit. It is often used to rehearse the operation plan (OPLAN) or OPORD when time is not an immediate constraint. One must plan extensively to execute full-scale rehearsals correctly and in a manner that does not waste the soldiers' time.

## *INITIATION OF PCCs, PCIs, AND MISSION-SPECIFIC DRILLS*

The subordinate leader conducts PCCs to determine his unit's readiness to execute its assigned tactical mission. These inspections are informal. They are the leader's inspection of mission-critical tasks or areas of special interest to the commander. PCCs should cover—

- Vehicles.
- Weapons systems.

- Soldier's equipment, to include his weapon.
- Mission-critical equipment, such as explosives, detonators, and blasting machines.

The unit commander or leader conducts PCIs to determine the unit's readiness to execute its assigned tactical mission. These inspections may be formal or informal. Formal PCIs are the commander's meticulous and time-consuming inspections of all areas within the unit. They are seldom conducted during combat operations or assembly-area occupation. Informal PCIs are the commander's inspections of particular areas or activities of special interest or concern to him. He may designate the executive officer (XO) or first sergeant (1SG) to perform the PCI. PCIs should cover—

- Vehicles
- Major weapons systems.
- Soldiers.
- Communications equipment.

Mission-specific drills are rehearsals of special combat drills that subordinate units conduct. Special drills may include—

- Breach drills.
- Marking drills.
- Situational obstacles.
- Reserve demolitions.

# CHAPTER 3
# Mobility and Survivability

Mobility and survivability are the primary tasks of the engineer on the battlefield. The platoon leader requires a detailed understanding of this BOS to succeed. This chapter provides the tactics, techniques, and procedures (TTP) needed for the platoon to conduct mobility, countermobility, and survivability.

## MOBILITY

Mobility enables the commander to maneuver tactical units into positions of advantage over the enemy. In the attack, engineers reduce enemy obstacles and fortifications that inhibit maneuver.

Expanded engineer reconnaissance capability is needed to identify routes, existing obstacles, and minefields. Open areas that are level provide good sites for aircraft landing strips; however, in most cases, the soil must be stabilized.

The application of the breaching fundamentals (suppress, obscure, secure, and reduce {SOSR}) and the organization of the force in terms of support, breach, and assault forces are standard. However, open areas may offer greater opportunity to bypass enemy obstacles because of the greater range of mobility afforded by the terrain. Exercise caution when choosing to bypass enemy obstacles, since the bypass may lead the force to the enemy's engagement area (EA). Additionally, expect the enemy to make extensive use of rapid mine-laying techniques that include SCATMINE delivery systems, as well as more conventional methods of mine emplacement.

### *OBSTACLE BREACHING*

The platoon leader must decide where the best breach location is. Never breach where the avenue of approach crosses the obstacle. The enemy is sure to have preplanned artillery there. Pick a location away from obvious breaching sites to reduce the threat of artillery. The platoon leader should place himself where he can see the breach site and effectively command and control his platoon. He must be able to signal his breaching reserve (generally a squad) if he sees the breach failing. Observation of the breach site also reduces reporting requirements on the squad leaders to breach and mark the breach site.

Typically, there is a time lag as the maneuver force suppresses the enemy and obscures the breach site for the breach. The platoon leader should also consider dismounting an element from his lead squad to reconnoiter the best breach location during this tactical pause. This will assist the platoon leader in choosing the best breach site.

Obstacle breaching must be a CA effort. Units are more successful if they conduct—

- Obstacle reconnaissance.
- CA rehearsals.
- Breach planning.

For more information on obstacle breaching, see *FM 90-13-1.*

### Breaching Fundamentals

When breaching against a defending enemy, the following fundamentals must be applied to ensure success:

- Suppress. Neutralize or destroy the enemy's weapons with our own fires. (The support force should have a 3:1 advantage over the enemy's weapons overwatching the obstacle.) Consider the effects and characteristics of the enemy's weapons. Ensure the correct focus of direct and indirect fires and the timely shift or lift of fires. Consider what ammunition is required.
- Obscure. Hamper the enemy's observation and target acquisition through the use of smoke (indirect, on board, pyrotechnic), terrain, limited visibility, and stealth.
- Secure. Eliminate enemy interference with obstacle reduction and lane usage. Occupy the obstacle with a friendly force and through the use of effective fires.
- Reduce. Create lanes through, over, or around an obstacle. Ensure that a company/team has at least one lane through an obstacle and that a battalion TF has two lanes. Consider what additional requirements are needed to guide follow-on forces through the obstacle or to mark and reduce the obstacle.

### Types of Breaching Operations

When planning a breaching operation, the engineer leader should consider bypassing the obstacle as his first option. However, the enemy's intent may be to force the unit to turn or channelize into a fire sac. This is why a good reconnaissance of a possible bypass is essential. The unit conducting a bypass must have maximum security during the bypass. If bypass operations are not possible, consider the following options:

***In-stride Breach***. A unit conducts an in-stride breach when it can be successfully executed under the complete $C^2$ of a task-organized subordinate unit or when the situation is unclear and rapid movement is critical. Subordinate units serve as both the breach and assault forces. The subordinate commanders synchronize the breaching fundamentals. Doctrinally, TFs and above conduct in-stride breaching.

***Deliberate Breach***. A unit conducts a deliberate breach when the force-allocation ratios for support, breach, and assault forces are beyond the capability of a task-organized subordinate unit. One or more subordinate units are specifically tasked to perform the role of support, breach, and assault forces. Synchronization is critical and the principle of mass drives task organization.

***Assault Breach***. A unit conducts an assault breach during the penetration and destruction phases of actions on the objective to destroy an enemy that has had time to emplace protective obstacles. For the assault breach, a rule of thumb is one lane per assaulting platoon. Assault lanes are generally only footpaths until the objective is secured. Mounted assault lanes require the same effort as tactical vehicle lanes.

***Covert Breach***. A unit conducts a covert breach during limited visibility when surprise is essential. The covert breach relies on stealth; quiet, manual lane-reduction techniques; and dismounted maneuver. It can be employed to breach tactical obstacles or protective obstacles (more difficult). It requires the same combat ratios as other breaches for suppression and security, and these points should be considered during the planning process.

### Breach Planning

The following are considered when developing a breaching plan (see *Figures 3-1 and 3-2, pages 3-4 and 3-5*):

- Reverse planning begins with actions on the objective.
- Actions on the objective drive the size of the assault force and determine the number and location of lanes to be breached.
- Lane requirements and the type of obstacle drive the allocation of mobility assets to the breach force.
- Ability of the enemy's infantry to interfere with the breach determines whether to secure the breaching site by force or by fire.
- Ability of the enemy to mass fires at the breaching site determines the amount of suppression required and the size of the support force.

The platoon leader must plan for actions after the breach is complete. Generally, the platoon is responsible for beaching, marking, and providing guides for the lane.

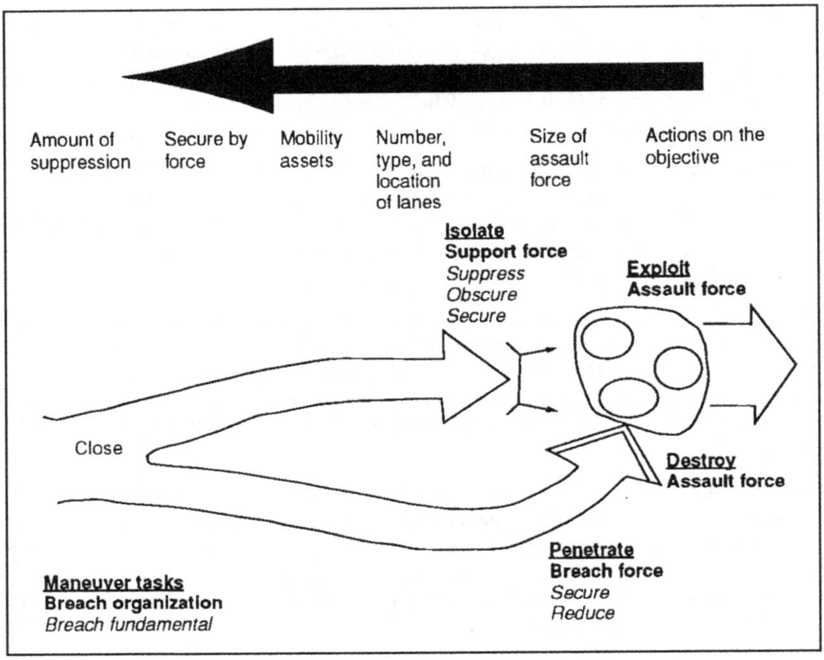


**Figure 3-1. Breach planning**

Normally, the platoon remains at the breach location and continues to improve the breached lane. The platoon must be prepared to reopen the lane if the enemy closes it with artillery-delivered mines. The platoon leader should also plan for vehicles that are stopped in the lane. The platoon may be forced to create an additional lane around the stopped vehicle or extract it from the breached lane.

### Breach TF Organization

The breach TF is organized into three elements:

- Support force—eliminates enemy interference with the breach, isolates the objective area, and destroys enemy weapons that are able to fire on the breach force.

- Breach force—creates lanes in and through the obstacle. While creating the lane through the obstacle, it must also locally secure the breaching site and pass the assault force through the breach.

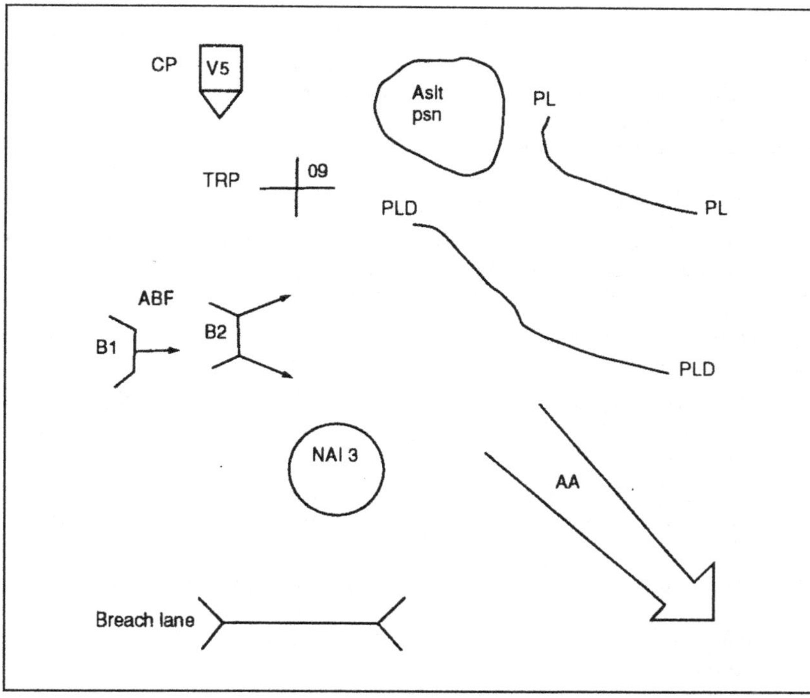


**Figure 3-2. Breach control measures**

- Assault force—destroys or dislodges the enemy force on the far side of the obstacle and seizes the initial foothold on the objective.

### Obscurants

To obscure friendly activities and movement, units can use artillery- or mortar-delivered smoke and/or white phosphorous (WP) rounds, generators, smoke pots and M203 grenades, chemical units, and burning diesel. Obscurants—

- Should be preplanned and coordinated with the company commander and the supported TF.
- Can be used for deception.
- Can be combined with supporting fires.
- Can be used over large areas.

### *LANE MARKING*

The critical components to any lane-marking system are the lane-marking patterns and devices. Lane-marking devices are not standardized Army-wide. *Figure 3-3* shows examples of lane-marking devices. *Table 3-1, page 3-8,* can be used as a guide to choose lane-marking devices. Lane-marking patterns have been standardized Army-wide. Standard breach lane-marking patterns are explained in *FM 90-13-1, Appendix E*. The following are lane markers and their use:

- Entrance markers—indicate the start and the width of a reduced lane. They must be visually different than handrail markers. Entrance markers are placed a minimum of 4.5 meters apart (1 meter for dismounted).
- Handrail markers—define the lane path and indicate the limits of the lane width. Mark, as a minimum, the left handrail.
- Exit markers—indicate the far-side limit of reduced lanes. Exit markers must be visually different from handrail markers but may be the same as entrance markers.
- Entrance-funnel markers—augment the entrance markers. They assist the small-unit commander in guiding the lead unit of his combat column formation.
- Final-approach marker—is a highly visible marker that augments the visual signature of entrance-funnel markers. It provides the assault-force commander with a highly visible reference point toward which to maneuver his formation.
- Far-recognition markers—are highly visible markers located between the final-approach marker and the friendly unit. They are primarily used when passing battalion-sized forces through a lane where distance, visibility, or terrain does not allow the passing force direct observation of the final-approach marker. When possible, far-recognition markers should be different from the final- approach marker.
- Traffic-control post (TCP) or guides—are a two-man team with communications means that assists the commander in controlling the movement of forces. When possible, military police (MP) should man TCPs. However, the breach force should plan to man TCPs until relieved.

The standard levels of lane marking for breach lanes and bypasses are initial, intermediate, and full (see *Figure 3-4, page 3-9*). The following are lane-marking patterns and their use:

- Initial lane-marking pattern—is emplaced by the breach force immediately after the lane is reduced and proofed. It is a signal to the assault force that the lane is ready for traffic to pass through.

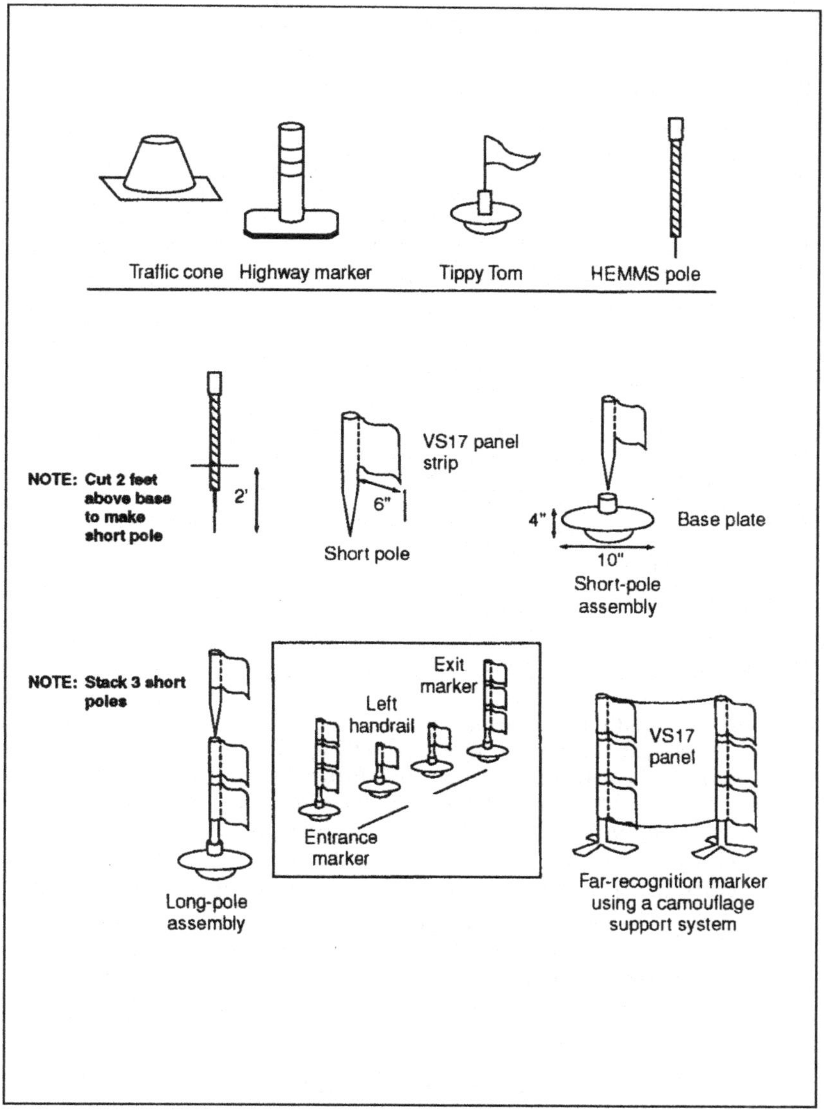


**Figure 3-3. Lane-marking devices**

Table 3-1. Breach lane marking

| Markers | Mounted Forces | Dismounted Forces |
|---|---|---|
| Handrail and funnel | They—<br>• Can be seen by the vehicle commander and driver buttoned up from 50 meters.<br>• Are quick to emplace, minimizing soldier exposure. | They—<br>• Are visible to a prone dismounted soldier from 50 meters.<br>• Are lightweight and easy to emplace. |
| Entrance and exit | They—<br>• Can be seen from 100 meters.<br>• Are visually different from handrail and funnel markers.<br>• Are easy to emplace and are man portable. | They—<br>• Are visible from 50 meters.<br>• Are visually different from handrail and funnel markers.<br>• Are easy to emplace and are man portable. |
| Final approach and far recognition | They—<br>• Can be seen from 500 meters.<br>• Are visually different from each other.<br>• Can be altered to facilitate traffic control through multiple lanes. | They—<br>• Are visible from 100 meters.<br>• Are visually different from each other.<br>• Can be altered to facilitate traffic control through multiple lanes. |

- Intermediate lane-marking pattern—is used for the commitment of larger combat forces who are unable to directly observe the breach or the rearward passage of sustainment traffic. It builds on the initial lane-marking pattern by adding right handrail markers, exit-funnel markers, far-recognition markers, and a far-side final-approach marker.
- Full lane-marking pattern—is usually not part of an initial breaching operation. It is a lane that will support uninterrupted two-way traffic.

*Table 3-2, page 3-10,* provides information as to who marks the various types of lanes, when they are to mark the lanes, and what markers they are to use. Lane marking is normally synchronized during the CA rehearsal to ensure that all key leaders understand the marking procedure and standard being employed.

## COUNTERMOBILITY

Countermobility operations are conducted to augment natural terrain with obstacle systems according to the commander's concept. This adds depth to the battle in

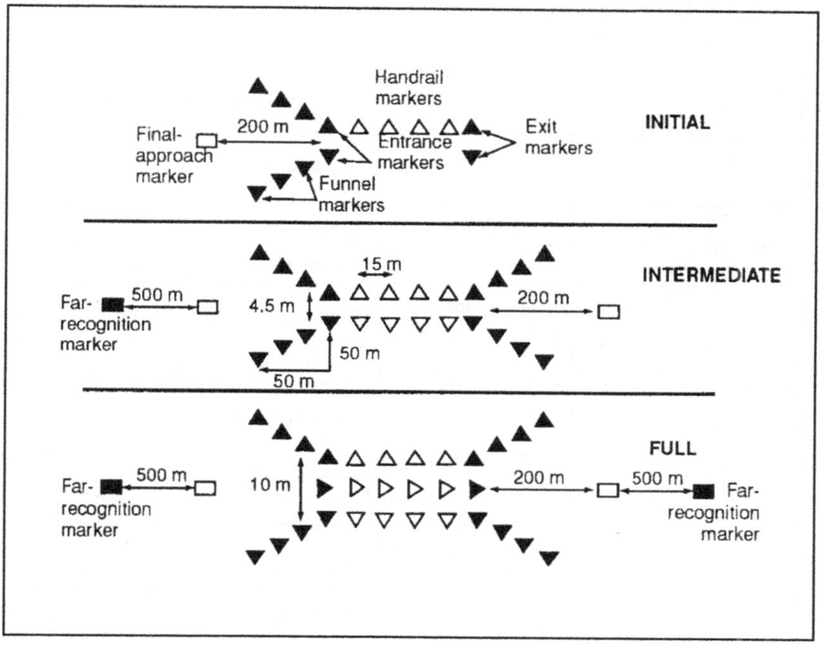


**Figure 3-4. Lane-marking patterns**

space and time by attacking the enemy's ability to maneuver its forces. With the enemy's movement impeded (disrupted, turned, fixed, or blocked), it is vulnerable to our forces. Engineers advise the commander on the best means to reinforce natural obstacles (terrain) and to emplace tactical obstacles that support his plan. Minefields, wire, antitank ditches (ATDs) or berms, road craters, bridge demolition, and all other traditional obstacles may be effective. For expedient obstacles, units should consider employing local materials.

ATD or berms require extensive preparation but are very effective when properly prepared. Many areas have irrigation ditches that can be used tactically. However, in sandy areas, ditches can easily be filled in. Obstacles should be planned to protect the unit's flanks during offensive operations.

All obstacles can be situational obstacles. Situational obstacles are obstacle resources held in reserve. They have a "be-prepared" mission but not an "on-order mission". Situational obstacles should be considered during planning to give the maneuver commander a flexible response to enemy or friendly maneuver.

**Table 3-2. Lane-marking level—unit responsible, trigger events, and lane markers**

| Breach Type | Initial | Intermediate | Full |
|---|---|---|---|
| | **Unit Responsible** | | |
| Deliberate | TF breach force | TF breach force | Brigade |
| Covert | TF breach force | TF breach force | Brigade |
| In-stride | Breach company/ team | TF mobility reserve | Brigade |
| Assault | Assault platoon | TF assault force | NA |
| | **Trigger Events** | | |
| | When—<br>• Lanes are reduced<br>• Passing platoon- or company-sized forces | When passing—<br>• Battalion- or company-sized forces<br>• Forces that cannot see the lane<br>• TF combat trains | When—<br>• Passing brigade- or battalion-sized forces<br>• Situation requires uninterrupted sustainment traffic |
| | **Lane Markers** | | |
| | Entrance/exit<br>Left handrail<br>Entrance funnel<br>Final approach | The following markers are added:<br>• Right handrail.<br>• Exit funnel.<br>• Far-side final approach.<br>• Far recognition.<br>• Guides/TCPs. | Lane width is expanded to 10 meters. Existing markers are adjusted. Far-side-recognition markers and guides/TCPs are added. |

## *OBSTACLE PLANNING*

To assist the TF commander in the defense, the TF engineer must integrate obstacle planning throughout the TF sector and coordinate with adjacent TFs according to the overall engineer plan.

Each echelon has its own obstacle-control measures for directing the location of obstacles on the ground. The division commander and his staff use obstacle zones to control obstacles to support future operations. Within these zones, the brigades plan obstacle belts, and the battalion TF plans obstacle groups. The engineer will then emplace each obstacle by doctrinal obstacle norms to support the commander.

### Protective Obstacles

They are designed to repel the enemy's assault by emplacing them around a unit's defensive position (a survivability task).

### Tactical Obstacles

They are designed and emplaced to (see *Figure 3-5*)—

- Disrupt, which—
  - Breaks up C[2].
  - Frustrates the enemy's timing.
  - Interrupts formations or attacks the enemy when they are massing into formations.
  - Coerces the enemy into our lines of fire.

| Obstacle-Effect Graphic | Application | Examples Conveying Intent | Resource Factor[1] |
|---|---|---|---|
| Disrupt 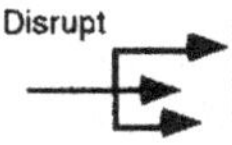 | Short arrow indicates where enemy is attacked by obstacles. Long arrows indicate where bypass is allowed and attacked by fires. | 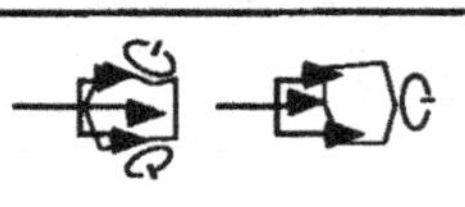 | 0.5 |
| Turn  | Heel of arrow is anchor point. Direction of arrow indicates desired direction of turn. | 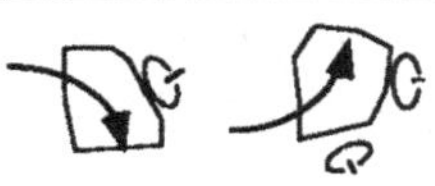 | 1.2 |
| Fix 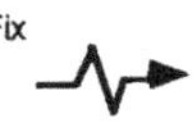 | Irregular part of arrow indicates where enemy advance is slowed by obstacles. | 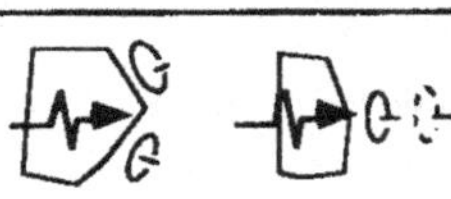 | 1.0 |
| Block  | The ends of the vertical line indicate the limit of enemy advance. The ends of the vertical line also indicate where obstacles tie in to severely restricted terrain. |  | 2.4 |

[1] See *FM 90-7, Appendix C* for an explanation of how to use the resource factor.

Direction of Enemy Attack →

**Figure 3-5. Obstacle-effect graphics**

- Fix, which—
  - Slows the enemy's advance within a specified area.
  - Causes the enemy to breach repeatedly.
- Turn, which manipulates enemy maneuver in a desired direction.
- Block, which—
  - Stops the enemy along a particular avenue of approach.
  - Defeats the enemy's breaching effort.

*Figure 3-6* shows how to employ conventional row minefields to obtain the desired obstacle effects and the planning figures for standard minefields by obstacle effect.

### Situational Obstacles

Situational obstacles are "triggered" by an event or situation. Once the general location of the obstacle has been determined, it must be refined to better enhance its effect on the enemy. When planning, preparing, and executing situational obstacles (see *FM 90-7*)—

- Identify the need.
- Plan for appropriate resources.
- Plan the obstacle.
- Integrate the obstacle with friendly fires.
- Identify obstacle execution triggers.
- Withhold execution of the obstacle until it is needed.

Situational obstacles can be used to—

- Attack an enemy's vulnerability.
- Exploit success.
- Separate follow-on enemy forces.
- Provide flank protection.

*Tables 3-3 and 3-4, pages 3-14 and 3-15,* provide some characteristics of conventional and scatterable mines and are useful in planning minefields. When emplacing standard disrupt, fix, turn, and block minefields, see *Figures 3-7 through 3-12, pages 3-16 through 3-21, and Tables 3-5 through 3-8, pages 3-22 and 3-23,* for information on the employment of SCATMINE systems. FASCAM systems include the air/ground Volcano, Modular Pack Mine System (MOPMS), Flipper, area denial artillery munition (ADAM), and remote antiarmor mine (RAAM).

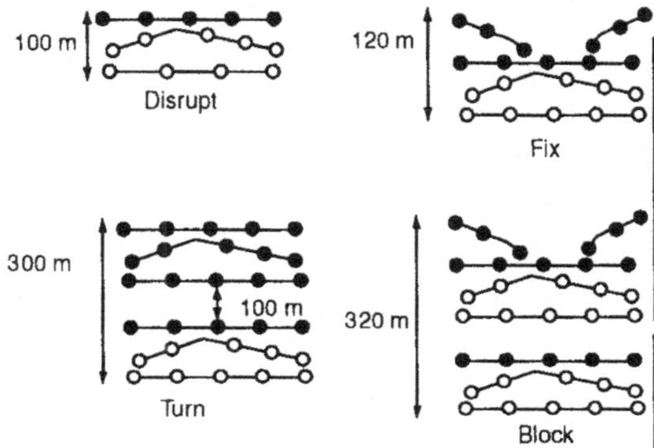


Mines spaced 6 m apart
Rows spaced 50 m apart
Minefield frontage
Disrupt: 250 m
Turn: 500 m
Fix: 250 m
Block: 500 m
o Track-width fuzed mine
● Full-width fuzed mine

AP mine employment

AP mines are—

- Used in block minefields only.
- Used in any two rows of full-width AT mines.
- Patterned based on AP mine type.

| | Disrupt | Turn | Fix | Block |
|---|---|---|---|---|
| Frontage (m) | 250 | 500 | 250 | 500 |
| Depth (m) | 100 | 300 | 120 | 320 |
| AT full (# rows) | 1 | 4 | 1 | 4 |
| AT track (# rows) | 2 | 2 | 2 | 2 |
| AP (# rows) | 0 | 0 | 0 | 2 |
| IOE (yes or no) | No | No | Yes | Yes |
| AHD (yes or no) | Yes | No | No | Yes |
| Plt hrs reqd | 1.5 | 3.5 | 1.5 | 5.0 |
| Track-width AT mines | 84 | 168 | 84 | 168 |
| Full-width AT mines<br>AP mines M16<br>AP mines M14 | 42<br>0<br>0 | 336<br>0<br>0 | 63<br>0<br>0 | 378<br>84<br>500 |
| Density AT | 0.5 | 1.0 | 0.6 | 1.1 |
| Density AP M16/M14 | 0 | 0 | 0 | 0.17/1.0 |

**Figure 3-6. Minefield norms**

Table 3-3. Conventional mines

| Mine | Arming Time | Fuzing | Warhead | AHD | Mine Weight | Mines per 5-Ton Dump | Remarks |
|---|---|---|---|---|---|---|---|
| M14 | NA | Pressure | Blast | No | 3.3 oz | 6,480 | |
| M16A1 | NA | Pressure Trip wire | Bounding fragment | Yes | 8 lb | 672 | |
| M18A1 | NA | | Directional fragment | No | 3.5 lb | 1,782 | |
| M86 (PDM) | 50 sec | | Bounding fragment | 100% | 1 lb | — | Self-destruct after 4 hours |
| M15 | NA | Pressure | Blast | Yes | 30 lb | 90 | |
| M15 w/ M624 | NA | Tilt rod | Blast | Yes | 30 lb | 90 | |
| M21 | NA | Tilt rod | Shape charge | Yes | 17 lb | 192 | |

## OBSTACLE SITING

When determining where to emplace obstacles, the platoon leader should consider the following:

- Supporting the maneuver commander's intent.
- Supporting the CATK plan.
- Supporting the defensive plan.
- Coordinating obstacle locations with the maneuver element, the fire-support officer (FSO), the Operations and Training Officer (US Army) (S3), and the weapons-systems commander.
- Prioritizing the engineer effort.
- Planning the maneuver unit's responsibility.
- Using observation and fields of fire, cover and concealment, obstacles, key terrain, and avenues of approach (OCOKA).

### Company/Team Commanders' Coordination

The platoon leader must ensure that key obstacles are covered by indirect fire. These obstacles become priority targets for the artillery. When coordinating with company/team commanders, the platoon leader should consider the following:

- Determining the obstacle intent.
- Determining the company's/team's tactical purpose.

Table 3-4. Scatterable mines

| Mine | Delivery System | Arming Time | Fuzing | Warhead | AHD | Self-Destruct Time | Mine Weight | Mines per 5-Ton Dump |
|---|---|---|---|---|---|---|---|---|
| M73 | 155-mm arty (RAAM) | 45 sec<br>2 min | Magnetic | M-S plate | 20% | 48 hr | 3.8 lb | NA |
| M70 | 155-mm arty (RAAM) | 45 sec<br>2 min | Magnetic | M-S plate | 20% | 4 hr | 3.8 lb | NA |
| M75 | GEMSS Flipper | 45 min | Magnetic | M-S plate | 20% | 5 days<br>15 days | 3.8 lb | 1,600 |
| BLU 91/B | USAF (Gator) | 2 min | Magnetic | M-S plate | No | 4 hr<br>48 hr<br>15 days | 3.8 lb | NA |
| M76 | MOPMS | 2 min | Magnetic | M-S plate | No | 4 hr (recycle 3 times) | 3.8 lb | 30 modules (510 mines) |
| Volcano | Ground/air | 2 min | Magnetic | M-S plate | No | 4 hr<br>48 hr | 3.8 lb | 160 canisters (800 mine) |
| M72 | 155-mm arty (ADAM) | 45 sec<br>2 min | Trip wire | Bounding fragment | 20% | 15 days<br>48 hr | 1.2 lb | 36 per M692 projectile |
| M67 | 155-mm arty (ADAM) | 45 sec<br>2 min | Trip wire | Bounding fragment | 20% | 4 hr | 1.2 lb | 38 per M731 projectile |
| M74 | GEMSS Flipper | 45 min | Trip wire | Blast fragment | 20% | 5 days<br>15 days | 3.2 lb | 1,600 |
| BLU 92/B | USAF (Gator) | 2 min | Trip wire | Blast fragment | 100% | 4 hr<br>48 hr<br>15 days | 3.2 lb | NA |
| M77 | MOPMS | 2 min | Trip wire | Blast fragment | 100% | 4 hr (recycle up to 15 days) | 3.2 lb | 30 modules (120 mines) |
| Volcano | Ground/air | 2 min | Trip wire | Blast fragment | 100% | 4 hr<br>48 hr<br>15 days | 3.2 lb | 160 canister (160 mines) |

- Ensuring that the obstacles the maneuver company/team commanders emplace on the battlefield serve as a combat multiplier and enhance the direct-fire plan.
- Providing forward security during obstacle emplacement.
- Conducting obstacle turnover with the maneuver company/team and ensuring that indirect-fire support is provided after the obstacle is finished.
- Rehearsing lane-closure plans.

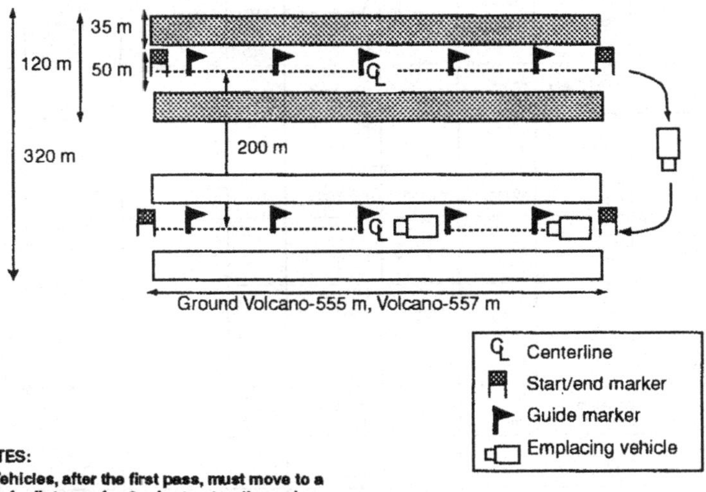


**NOTES:**

- **Vehicles, after the first pass, must move to a safe distance for 2 minutes to allow mines failing the arming sequence to self-destruct.**
- **Guide markers are offset from centerline.**
- **Start and end markers should be used for air Volcano only.**
- **Turn and block minefields require the dispensing of 80 canisters (40 each side) on each pass, 160 canisters total.**
- **One full load of ground or air Volcano emplaces one turn or block minefield.**

**Figure 3-7. Volcano turn and block minefields**

### Direct-Fire Integration

When integrating obstacles with direct fire, the platoon leader should consider the following:

- Coordinating with the maneuver element overwatching the obstacle.
- Sighting obstacles with weapons.
- Coordinating with the maneuver element for—
  - Forward security.

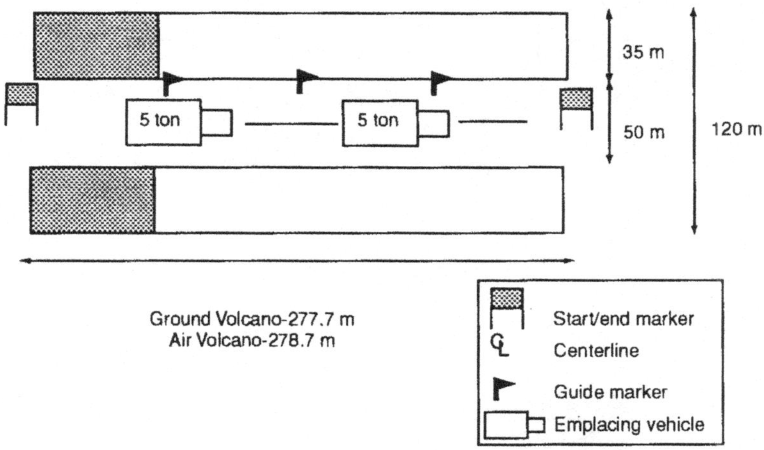


**Notes:**

- **Guide markers are offset to the left of the centerline to guide the driver and allow the vehicle to remain on the centerline.**
- **Start and end markers should be used for air Volcanos only. Guide markers are not required.**
- **Disrupt and fix minefields require the dispensing of 40 canisters (20 each side).**
- **One full load of air or ground Volcano provides four disrupt or fix minefields.**

**Figure 3-8. Volcano disrupt and fix minefields**

- Logistics support.
- Obstacle-turnover requirements.

### Indirect-Fire Integration

The platoon leader must ensure that key obstacles are covered by indirect fire. These obstacles become priority targets for the artillery. The platoon leader coordinates indirect fire with the company FIST and company/team commander in heavy units. Indirect fire must be coordinated with the FSO and the TF commander if units are light or if the platoon leader is the TF engineer. For more information on

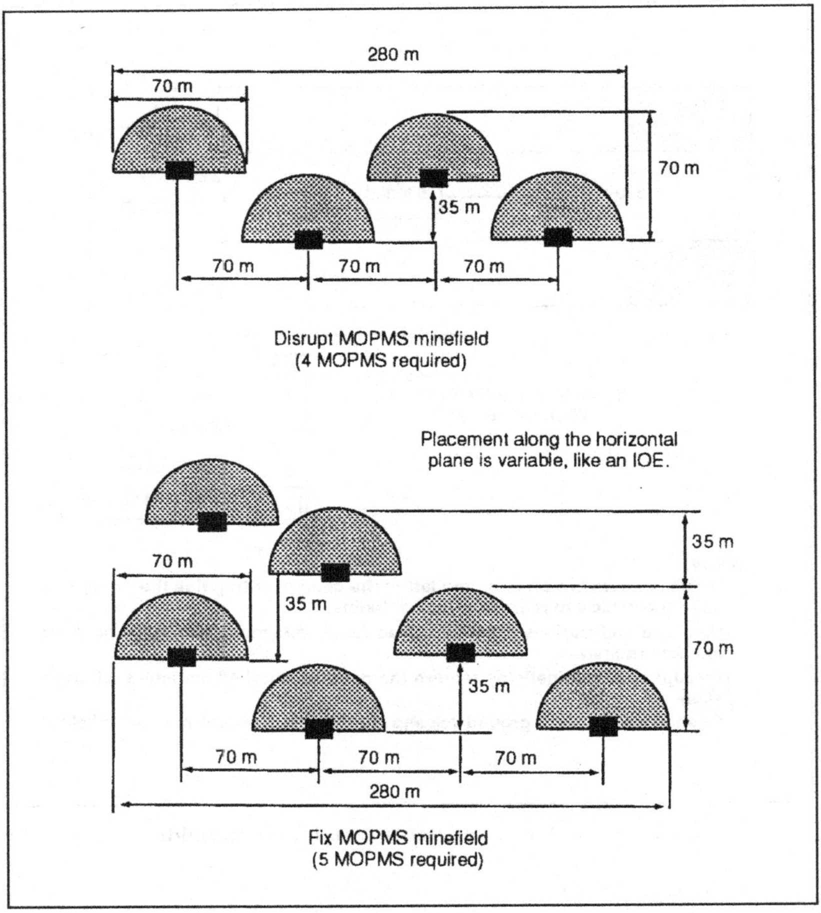


**Figure 3-9. MOPMS disrupt and fix minefields**

integrating obstacles with indirect fire, see *FM 90-7*. When integrating obstacles with indirect fire, the platoon leader should consider the following:

- Ensuring that there are target reference points (TRPs) on every obstacle.
- Preplanning ADAM and RAAM fires.
- Calling for fire nets (engineers).

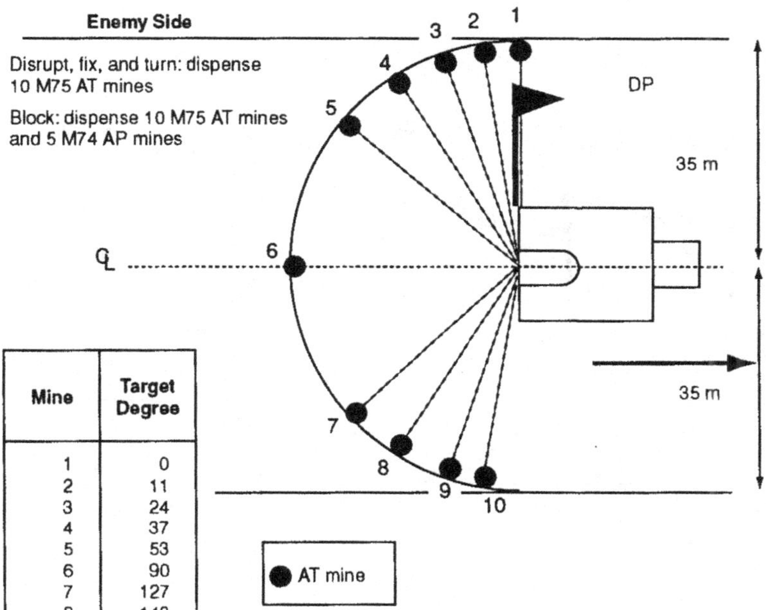


| Mine | Target Degree |
|---|---|
| 1 | 0 |
| 2 | 11 |
| 3 | 24 |
| 4 | 37 |
| 5 | 53 |
| 6 | 90 |
| 7 | 127 |
| 8 | 143 |
| 9 | 156 |
| 10 | 168 |

**Figure 3-10. Flipper TTP**

### *OBSTACLE CAMOUFLAGE (IF REQUIRED)*

When camouflaging obstacles, the following should be considered:

- Ensuring that the obstacles emplaced support the intent (sometimes you want the enemy to see the obstacles).
- Using reverse slope siting.
- Burying mines.
- Using night and covert emplacement.
- Minimizing dust signatures from the excavation.

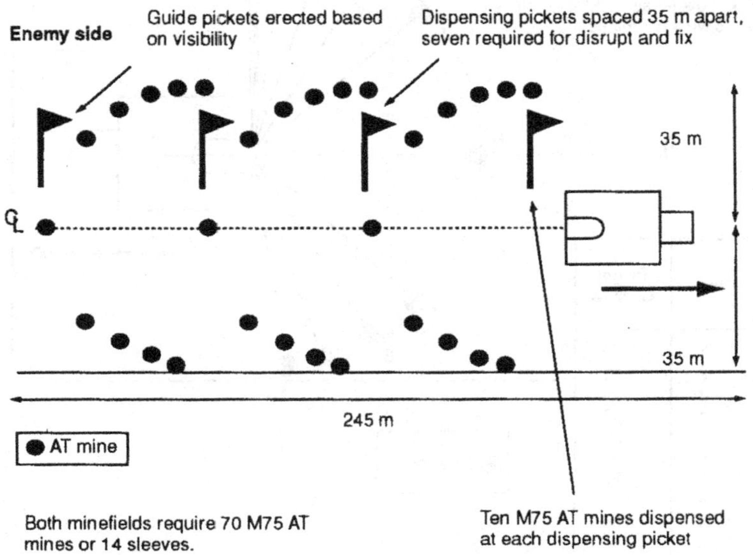


**Figure 3-11. Flipper fix and disrupt minefields**

### *OBSTACLE PROTECTION*

The following should be considered for the protection of obstacles:

- Denying the enemy covert breaching opportunity.
- Denying the enemy reconnaissance
- Using observation posts (OPs) and counterreconnaissance.
- Displacing forces during low visibility to cover obstacles.

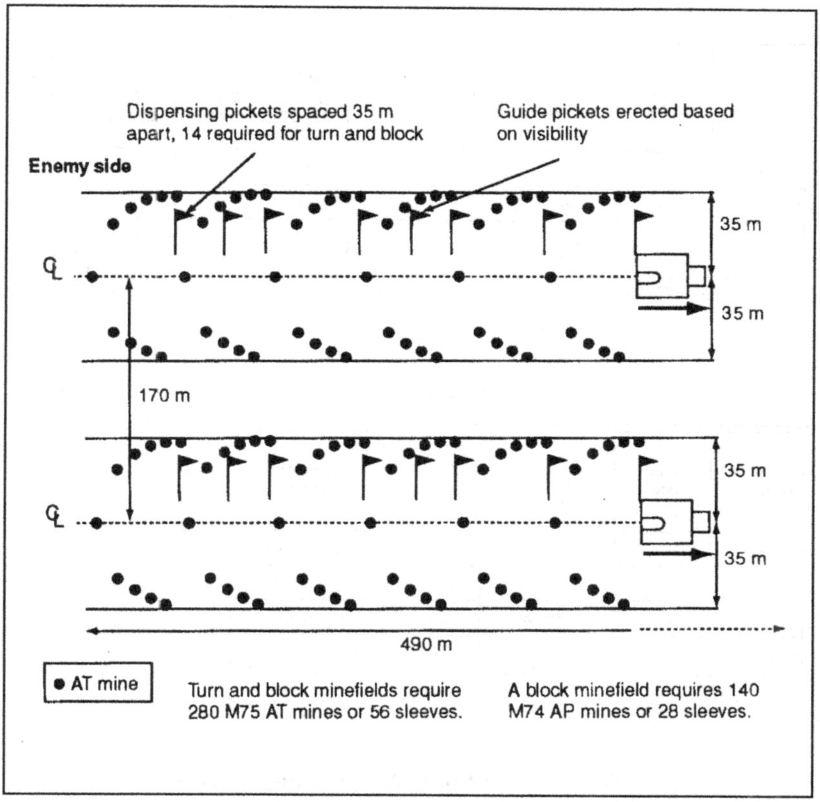


**Figure 3-12. Flipper turn and block minefields**

- Identifying and destroying the enemy's obstacle breaching equipment and $C^2$ vehicles.
- Covering the obstacle with direct and indirect fires.
- Patrolling obstacles at night.

## SURVIVABILITY

Preparation of fortifications may be difficult. Fortifications in sandy soils often require revetments. It may be impossible to dig in rocky areas, arctic regions, or locations with a high or perched water table. To counter these problems, you may

## Table 3-5. Flipper planning factors

| Type of Minefield | Depth (Meters) | Front (Meters) | Number of Strips | DPs per Strip | M75 Mines per DP | M74 Mines per DP | Total M75 Mines | Total M74 Mines |
|---|---|---|---|---|---|---|---|---|
| Disrupt | 70 | 245 | 1 | 7 | 10 | 0 | 70 | 0 |
| Fix | 70 | 245 | 1 | 7 | 10 | 0 | 70 | 0 |
| Turn | 240 | 490 | 2 | 14 | 10 | 0 | 280 | 0 |
| Block | 240 | 490 | 2 | 14 | 10 | 5 | 280 | 140 |

## Table 3-6. Volcano planning factors

| Type of Minefield | Depth (Meters) | Front Ground/Air (Meters) | Number of Strips | Canisters per Strip | Total Canisters | Minefields per Load |
|---|---|---|---|---|---|---|
| Disrupt | 120 | 277/278 | 1 | 40 (20 each side) | 40 | 4 |
| Fix | 120 | 277/278 | 1 | 40 (20 each side) | 40 | 4 |
| Turn | 320 | 555/557 | 2 | 80 (40 each side) | 160 | 1 |
| Block | 320 | 555/557 | 2 | 80 (40 each side) | 160 | 1 |

## Table 3-7. RAAM/ADAM planning factors

| Type of Minefield | RAAM | | ADAM | | Width (Meters) | Depth (Meters) |
|---|---|---|---|---|---|---|
| | Area | Linear | Area | Linear | | |
| Disrupt | 0.001 | 0.2 | 0.0005 | 0.1 | 200 | 200 |
| Turn | 0.002 | 0.9 | 0.001 | 0.4 | 400 | 400 |
| Fix | 0.002 | 0.4 | 0.0005 | 0.1 | 200 | 200 |
| Block | 0.004 | 1.6 | 0.002 | 0.8 | 400 | 400 |

**Table 3-8. Mine dump planning factors**

| Number of Personnel | Quantity of Mines |
|---|---|
| 2-man team (2 minutes per mine) | 25 mines per hour |
| Squad (8 soldiers) | 100 mines per hour |
| Platoon | 300 mines per hour; 3,000 mines per day |
| Company | 6,000 mines per day |
| ***NOTE:** For planning, soldiers work 10 hours per day.* | |

have to build up emplacements with existing materials or depressions. *FMs 5-34 and 5-103* provide the standards for dismounted and vehicular field fortifications.

Camouflage is very effective when properly employed. However, you must carefully select patterns and techniques to apply to the local environment. All equipment should have camouflage nets. Other survivability techniques include dispersion and frequent moves.

Direct-fire weapons are more effective from dug-in positions. Logistics areas (brigade support area/division support area (BSA/DSA)) require survivability support. These sites are large, relatively static, and difficult to camouflage. As a result, they are vulnerable to enemy interdiction. Emphasis should be placed on ammunition and fuel dumps.

Digging assets, such as bulldozers, small emplacement excavators (SEEs), or armored combat earthmovers, M9 (ACEs), should be tasked to provide survivability support to these sites. Particular emphasis should be placed on hardening ammunition and fuel-storage locations.

With long-range observation and fields of fire, defensive positions are very vulnerable to offensive fire. This fact, coupled with a lack of natural obstacles, may lead the commander to concentrate the bulk of his engineer effort on survivability positions. Survivability positions enhance the ability of all direct-fire elements to survive indirect-fire preparations and the direct fire from attacking enemy units.

Care should be taken when digging foxholes and tank hide positions in sandy or noncohesive soils since the walls have a tendency to cave in.

## *PLANNING*

Survivability positions are key to surviving any battle on today's modern battlefield. This is especially true where there is a lack of vegetation and natural hiding positions. The TF engineer must know the following information to start planning:

- Number of positions: primary, alternate, and supplementary.
- Type of position: hull-down, turret-down, or hasty.
- Weapons and vehicle priorities.
- Battle-position and EA priorities.
- Unit priorities.

Once the TF engineer determines the number of positions required, then the number of positions that can actually be put in is determined. The engineer unit's capability is based on the following factors:

- Ten-hour days for soldiers.
- Fifteen-hour days for equipment.
- Time available (days).
- Number of blade teams (BTs) available. ACEs, dozers, combat engineer vehicles (CEVs), and bucket loaders can be used to make blade teams. A blade team must include at least one dozer or ACE.
- Work rates.

For work rates, use the planning factor developed for your area of operation. If work rates are unknown, use the following as a guide:

- One hull-down position (HDP) takes 1.5 BT hours.
- One turret-down position (TDP) takes 3.5 BT hours.
- Fifty meters of ATD takes 1 BT hour.

A variety of variables may increase or decrease the time required to dig a vehicle fighting position or ATD. These variables include, but are not limited to, local soils and terrain, operator experience, additional maintenance down time, and the tactical situation. The difference of blade widths may also affect work rates. The M9 ACE has a shorter blade width than the D7 dozer and requires a minimum of two passes to cut the same width.

Survivability assets should be massed on the commander's number one priority. This focuses the survivability effort and allows the TF engineer to monitor construction progress. It also reduces the $C^2$ burden on the engineer platoon and facilitates rapid transition to changing priorities.

The following equations can be used to determine the number of HDPs and TDPs and the length of ATDs that can be emplaced:

HDP = (# BTs) x (1 HDP/work rate) x (# days) x (15 hrs/day)

TDP = (# BTs) x (1 TDP/work rate) x (# days) x (15 hrs/day)

Meters of ATD = (# BTs) x (ATD m/work rate) x (# days) x (15 hrs/day)

## *POSITIONS*

Coordination between the engineer/equipment operator and the maneuver element is critical to ensure the proper placement of positions. Without effective coordination, much valuable time is wasted redigging positions. The maneuver element, preferably the gunner or tank commander (TC), must tell the engineer/equipment operator the following:

- Type of position (deliberate or hasty).
- Gun target line.
- Sector of fire.
- TRP.

In heavy units, initial vehicle-fighting-position construction starts with the M9 ACE. The ACE has different digging capabilities than the dozer. In many cases, divisional ACEs start with hasty positions, which are then upgraded to improved positions as time allows. Also, the ACEs may start work before corps-level dozers arrive to augment the construction effort. In this situation, the ACEs would dig the hasty positions and the dozers would improve them. The platoon must plan for this augmentation to efficiently integrate corps assets into the construction effort. In heavy units, the company commander must integrate corps assets; however, in light units, the platoon leader must plan for and integrate corps assets.

*Figure 3-13, page 3-26,* shows the different vehicle fighting positions. Some terrain areas may not accommodate a hull or turret defilade. The use of a modified hasty vehicle-position design or simple marking techniques incorporated with terrain mottling may be required.

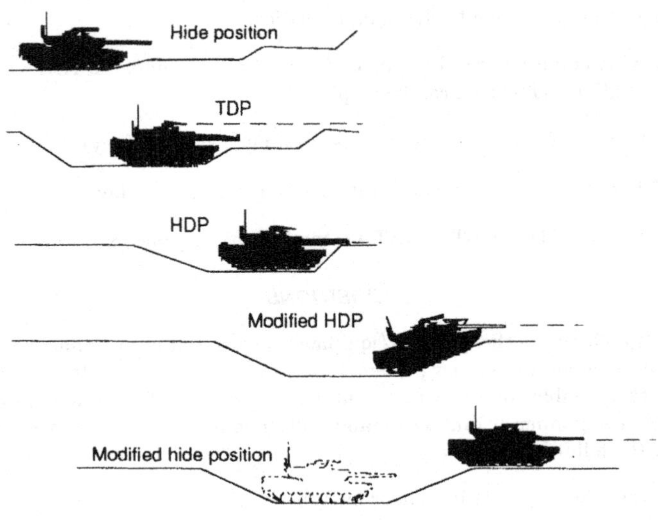


**Figure 3-13. Vehicle survivability positions**

Each soldier, not just engineers, must also emplace individual and crew fighting positions. Engineers have some limited assets, such as the SEE, that can assist the maneuver commander in accomplishing this mission. Fighting positions are—

- Located to effectively engage the enemy and mutually support each other.
- Designed to allow weapons to be properly fired.
- Constructed to provide adequate cover (see *Table 3-9*).
- Concealed from the enemy.

### *NUCLEAR, BIOLOGICAL, CHEMICAL (NBC)*

Chemical and biological weapons are many countries' preferred weapons of mass destruction. Many of our potential adversaries have an NBC capability. *Figure 3-14 and Tables 3-10 and 3-11, page 3-28,* detail the NBC threat and protective procedures.

## Table 3-9. Survivability

| Camouflage | Defensive Positions |
|---|---|
| Soldiers should consider the following when camouflaging:<br>• Breaking up the shape of the object.<br>• Shading the object with natural shadows.<br>• Concealing the shine of the object.<br>• Disrupting the silhouette of the object.<br>• Using folds, gullies, and wadies.<br>• Using reverse-slope positions.<br>• Scaring two to three positions for each vehicle or position.<br>• Using irregular-shaped positions<br>• Building positions a minimum of twice the size of the concealed object.<br>• Exercising noise and light discipline.<br>• Using camouflage for deception. | Soldiers should know and consider the following facts when constructing defensive positions:<br>• Field expedient construction materials, such as corrugated metal, coverts, scrap metal, 40-gallon oil drums, sandbags, and sand-filled ammo boxes, can be used.<br>• Small-arms fire penetrates loose sand 39 inches and sandbag parapets 24 inches.<br>• A minimum of 18 to 24 inches of overhead cover is required to protect against indirect fire.<br>• Revetments are required in sandy soils.<br>• Design details are covered in *FMs 5-34 and 5-103.* |

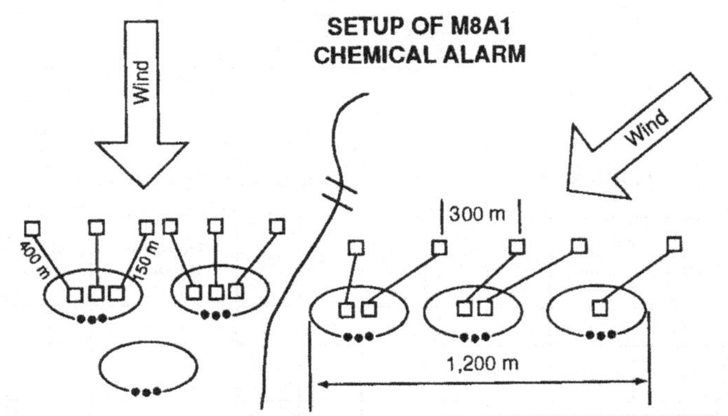


CAM put in operation

1. Remove nozzle's protective cap and place cap onto environmental cap.
2. Place filtered nozzle standoff onto nozzle (do not touch nozzle).
3. Insert nozzle in "H" end of confidence sample for 1 second.
   - Verify that at least 3 bars are displayed.
   - Ensure that the display has no more that 1 bar within 2 minutes.
4. Check in "H" and "G" modes.

**Figure 3-14. M8A1 chemical alarm**

## Table 3-10. Unmasking procedures

| Unmasking with M256 Series Kit | Unmasking without M256 Series Kit | MOPP Gear Exchange |
|---|---|---|
| This procedure takes 10 minutes. See *FM 3-4* for more information.<br>The leader—<br>• Checks M256 results to ensure that they are negative.<br>• Selects one or two soldiers.<br>The soldiers—<br>• Move to a shady place.<br>• Unmask for 5 minutes.<br>• Clear and reseal their masks.<br>The leader—<br>• Observes the soldiers for 10 minutes.<br>• Gives the all-clear signal if the soldiers show no symptoms.<br>• Watches for delayed symptoms.<br>• Has first aid available. | This procedure takes 25 minutes. See *FM 3-4* for more information.<br>The leader—<br>• Checks the M8 paper for liquid contamination.<br>• Selects one or two soldiers.<br>The soldiers—<br>• Move to a shady place.<br>• Break the seal of their mask, with their eyes open, for 15 seconds.<br>• Clear and reseal their masks.<br>The leader observes the soldiers for 10 minutes. If no symptoms occur, the soldiers—<br>• Unmask for 5 minutes.<br>• Clear and reseal their masks.<br>The leader—<br>• Observes the soldiers for 10 minutes.<br>• Gives the all-clear signal if the soldiers show no symptoms.<br>• Watches the soldiers for delayed symptoms.<br>• Has first aid available | Soldiers conduct MOPP gear exchange as a two-man team. See *FM 3-5* for more information.<br>Soldiers—<br>• Decon their individual gear.<br>• Prepare their mask for decon and their overgarment for removal.<br>• Decon their mask hood.<br>• Remove their overgarment and overshoes.<br>• Remove their gloves.<br>• Put on their new overgarment.<br>• Put on their new overshoes and gloves.<br>• Secure their hood.<br>• Secure their gear. |

## Table 3-11. NBC threat, detection, and protection

| Agent Type | Agent Class | Agent Effect | M8A | M256 | CAM | M8 | M9 | Protection Required | Skin Decon | Equip Decon |
|---|---|---|---|---|---|---|---|---|---|---|
| AC | NP | Blood | N | Y | N | N | N | Mask only | N | N |
| GA | NP | Nerve | Y | Y | Y | Y | Y | Mask and overgarment | Y | N |
| GB | NP | Nerve | Y | Y | Y | N | N | Mask and overgarment | Y | N |
| GD | P | Nerve | Y | Y | Y | Yellow | Y | Mask and overgarment | Y | Y |
| VX | P | Nerve | N | Y | N | Green | Y | Mask and overgarment | Y | Y |
| HL | P | Blister | N | Y | Y | Red | Y | Mask and overgarment | Y | Y |
| HD | P | Blister | N | Y | Y | Red | Y | Mask and overgarment | Y | Y |
| CG | NP | Choking | N | N | N | N | N | Mask only | N | N |

# CHAPTER 4
# Fire Support

The platoon gets FS in the forms of indirect fires, smoke, and obstacle emplacement. FS protects the platoon while it is executing its mission. The platoon may provide survivability to FS units. The platoon leader must—

- Keep the TOC apprised of all element locations at all times. The FSO gets information on engineer unit locations from the TOC operations section (there are no FISTs with engineers).
- Work closely with the FSO or the FIST to plan and report SCATMINEs, cover friendly obstacles with fire, and provide suppressive fires and smoke on locations of expected enemy obstacles.
- Keep track of the amount of artillery FASCAM allocated to the TF and still available for employment.
- Get the frequencies, call sign, and procedures needed to call for indirect fires for the platoon. Most call for fires will be done by adjusting fires from a target or known point. See *FM 5-34* for call-for-fire procedures. Two examples of call-for-fire procedures follow:
  - From your location, you will adjust fire by distance and direction from the TRP to the target. Give the distance in meters and the direction as you see it. If the target is to the left or right of the TRP as you see it, call LEFT distance in meters or RIGHT distance in meters. If the target is behind the TRP as you see it, call ADD distance in meters and if the target is in front of the TRP, call DROP distance in meters.
  - During breaching operations, you may have to call for smoke. Ensure that the FSO or the FIST knows the location and the duration of your smoked target. The call-for-fire procedures are similar to those discussed above.
- Determine if the maneuver commander plans to provide survivability positions for artillery assets and what the priority is for those positions.
- Ensure that obstacles are synchronized with the indirect-fire plan.

# CHAPTER 5
# Air Defense

The platoon leader and sergeant should understand how AD weapons can support the platoon and how the platoon can support AD units. The platoon leader should know where the AD umbrella is in relationship to his area of operations and what kind of support he can expect. The platoon can provide its own AD measures, either passive or active. Passive AD measures are used to keep soldiers and equipment undetected by enemy air assets. Active AD measures are used to engage enemy aircraft. The following are active AD measures:

- Using the football-field method for high-performance aircraft (aim 200 meters in front of the aircraft).
- Using the football-field method for helicopters (aim 50 meters in front of the helicopter).
- Aiming weapons slightly above the nose of aircraft that is approaching head on.
- Firing all weapons at the reference point
- Ensuring that the AD-weapons control status is given. Status could be—
  - Hold—weapons are fired only in self-defense or in response to a formal order.
  - Tight—weapons are fired only at aircraft positively identified as hostile.
  - Free—weapons are fired at any aircraft not positively identified as friendly.
- Ensuring that AD warning conditions are given. Conditions could be—
  - White—an air attack is not probable.
  - Yellow—an air attack is probable.
  - Red—an air attack is imminent or in progress.
- Determining if the maneuver commander plans to provide survivability positions for AD assets and what the priority is for those positions.
- Ensuring that the platoon is able to receive AD early warning. When the AD warning condition changes to red, the warning of incoming enemy aircraft and their direction is transmitted over command or operations and intelligence networks. The platoon leader should ensure that the platoon is monitoring the appropriate networks.

# CHAPTER 6
# Combat Service Support

The platoon's CSS is determined by its supply requirements. The platoon leader and sergeant decide what CSS the platoon needs for an operation and initiate a request. Their request is forwarded to the 1SG of either their parent engineer company or supported maneuver unit, depending on their command and support relationships. For more information on command and support relationships, see *Chapter 2*. The platoon's CSS is normally delivered in a scheduled logistical package (LOGPAC).

## RESUPPLY OPERATIONS

A LOGPAC is a resupply element that is formed on the basis of a unit's logistic requirements. Normally, it consists of a petroleum, oils, and lubricants (POL) truck; an ammunition truck (Class IV/Class V supplies); and a supply truck that carries rations, repair parts, and other requested items and pulls a water trailer. LOGPACs are—

- Assembled in the BSA and then moved forward under the control of the support-platoon leader to a logistics release point (LRP).
- Picked up by the 1SG at the LRP and moved forward to a secure area behind the company's position to feed, fuel, and resupply the company.
- Returned by the 1SG to the LRP where the support-platoon leader takes control of it and moves it back to the BSA.

LOGPACs are usually distributed by one of the following methods:

- Service-station method.
- Tailgate method.
- Modification of both.

In the service-station method, the 1SG establishes the LOGPAC in a centrally located and secure site. Successively, each platoon moves to the LOGPAC site. Each squad or crew passes through the various stations before the final inspection and returns to the work sites. Normally, the stations include maintenance support, fuel (POL), food, water, mail, administrative support, medics, and other CSS assets as required by the SOP and the mission, enemy, terrain, troops, and time available (METT-T). *Figure 6-1* shows the service-station method.

The tailgate method differs from the service-station method in two major aspects (see *Figure 6-2*). In the tailgate method, the LOGPACs—

- Are brought to each platoon's general location by the 1SG.
- Are usually not as comprehensive as the service-station LOGPACs because they are more mobile and spend more time in transit.

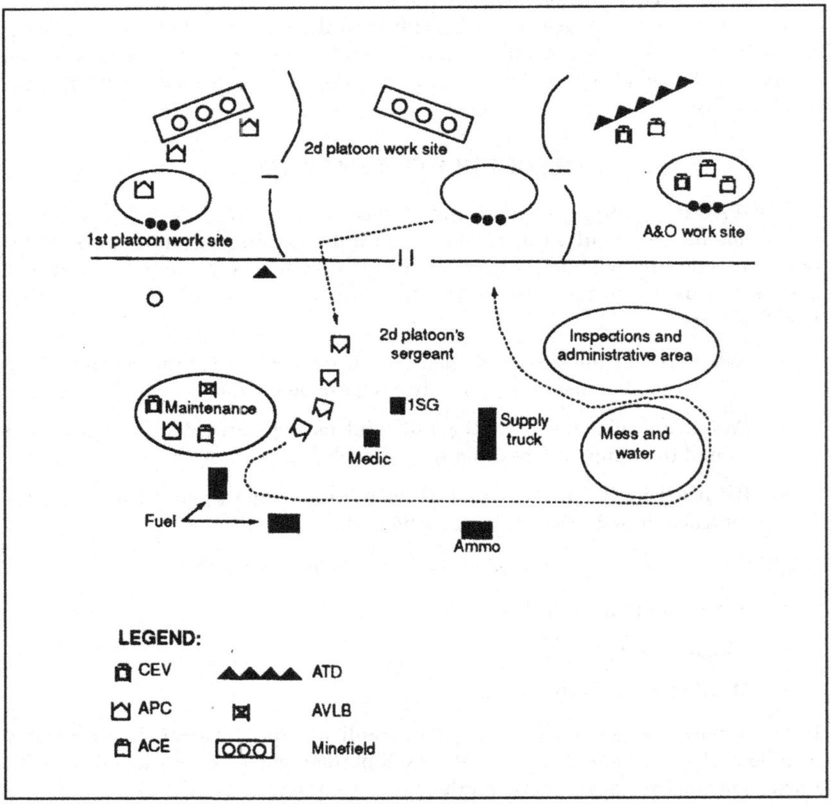


**Figure 6-1. Service-station method**

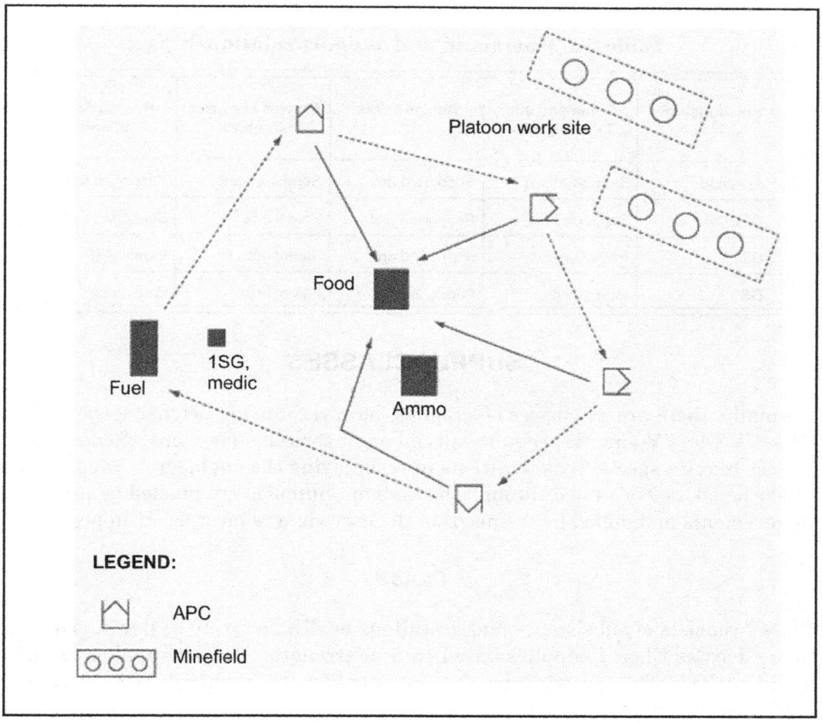


**Figure 6-2. Tailgate method**

The tailgate method takes more time for the 1SG but is the least interfering of the platoon's work effort. The platoon does not have to spend time moving to the LOGPAC.

Platoons often get resupplied from the maneuver company's/team's LOGPAC using the service-station or tailgate method. For security reasons, maneuver units normally provide LOGPACs to individual vehicles. Therefore, the squad rather than the platoon gets resupplied.

Even when being logistically supported by maneuver elements, platoons sometimes need engineer-specific resupply. Logistics-support responsibilities for command and support relationships are shown in *Table 6-1, page 6-4.*

**Table 6-1. Command and support relationships**

| Relationship | Command and Task Organize | Assign Tasks | Provide Logistics Support | Reorganize as Infantry |
|---|---|---|---|---|
| Attached | Supported unit | Supported unit | Supported unit | Supported unit |
| OPCON | Supported unit | Supported unit | Parent unit | Supported unit |
| DS | Parent unit | Supported unit | Parent unit | Parent unit |
| GS | Parent unit | Parent unit | Parent unit | Parent unit |

## SUPPLY CLASSES

Normally, there are 10 classes of supplies; however, during defensive operations, Class IV/Class V supplies require special engineer considerations. Some supply classes require special considerations in resupplying the engineers. Supplies can either be pushed or pulled through the system. Supplies are pushed by predicted requirements and pulled by the needs of the user via a request for resupply.

### *CLASS I*

Class I consists of subsistence and gratuitous health and welfare items. The platoon allocates Class I supplies based on unit strength. Class I supplies are normally pushed. The following are the amounts of water required for a soldier per day for—

- Drinking - 6 gallons.
- Personal hygiene - 2.5 gallons.
- Food preparation - 1.0 gallon.
- Vehicle radiators - 25 percent of the radiator's capacity.

### *CLASS II*

Class II consists of such items as clothing, individual equipment, tentage, hand tools, administrative and housekeeping supplies, chemical-defense and decontamination items, and wet- and cold-weather contingency items. The company allocates Class II supplies based on need. It is normally pulled.

### *CLASS III*

Class III consists of POL, including petroleum fuels, lubricants, hydraulic and insulating oils, preservatives, liquids and gases, bulk chemical products, coolants, deicer and antifreeze compounds, components and additives of petroleum and chemical products, and coal. Supply amounts are based on either offensive or defensive operations. When the TF is preparing to defend, additional Class III supplies are required for all equipment. ACEs require fuel every 8 hours when being heavily used. Additional fuel and package products, such as hydraulic fluid, should be maintained in the combat trains for this purpose. Forward fueling at the work sites should be a standard procedure. Class III supplies are usually pushed.

### *CLASS IV*

Class IV consists of construction materials, including all fortification and barrier materials. These are items for which allowances are not prescribed. Class IV supplies are pulled.

### *CLASS V*

Class V consists of all types of ammunition, including radiological and special weapons, bombs, explosives, mines, fuzes, detonators, pyrotechnics, missiles, rockets, propellants, and other associated items. Engineers need some unique Class V supplies (for example, cratering charges and line charges), which require special coordination with the Supply Officer (US Army) (S4) to get them into the TF's Class V system. Class V supplies are pulled.

### *CLASS VI*

Class VI consists of all personal-demand items, such as candy, cigarettes, soap, and cameras (nonmilitary sales items). Sundry packs are also Class VI items. The platoon allocates Class VI supplies based on need. Class VI supplies are normally pulled.

### *CLASS VII*

Class VII consists of major end items, such as launchers, tanks, mobile machine shops, vehicles, and organizational tool sets. Class VII supplies are demand driven.

### *CLASS VIII*

Class VIII consists of medical material, including repair parts peculiar to medical equipment. Class VIII supplies are normally pulled.

### *CLASS IX*

Class IX consists of repair parts and components, including kits, assemblies, and subassemblies (repairable and unrepairable) that are required for maintenance support of all equipment. The platoon's stock of repair parts is based on vehicle density, the authorized stockage list (ASL), the high-demand prescribed load list (PLL), and specific engineer-equipment requirements. Repair parts for items that the TF does not stock (for example, the M9 ACE or dozer) will have to come from the engineers. The platoon generates requests for repair parts by accurately completing *DA Form 2404*. When the maintenance contact team gets the form, it determines what repair parts are needed and the level of repair. The mechanics may bring the repair parts with them or the crew/operator may get them from the LOGPAC. Class V supplies are pulled.

### *CLASS X*

Class X consists of materials to support nonmilitary programs, such as agriculture and economic development. Class V supplies are pulled.

### *SPECIAL CLASS IV/CLASS V SUPPLIES*

These classes of supplies form the majority of the materials with which engineers construct the obstacles and fortifications required for the defense.

There are two types of Class IV/Class V loads: mission and basic. Mission loads consist of those materials required for a specific mission (for example, a standard fix minefield). Basic loads consist of those materials that the platoon carries to protect itself. The basic load can be used for missions to save time; however, it is to be replenished from the materials in the mission load. Basic loads are pulled and mission loads are pushed.

Mission loads are a TF responsibility regardless of the command and support relationships specified for engineers. The quantity of Class IV/Class V supplies (mines) normally stretches or exceeds the transportation assets of the TF. Palletized standard loads help solve the planning and distribution problem. Class IV/Class V supply points should be selected by their proximity to major obstacles.

Class IV/Class V resupply for the defense is one of the most demanding logistics operations the TF must carry out and requires all the assets that can be made available. A total cooperative effort by the TF, including engineers, is required if the defense is to be adequately resourced.

In the offense, TF planners anticipate Class IV/Class V mission loads for even a hasty defense on the objective. Prepackaging in the BSA speeds movement forward when it is needed. In addition to the Class V supplies that come forward as a part of the LOGPAC, the TF will maintain emergency Class V resupply in the combat trains.

## MAINTENANCE PROCEDURES

When the TF is providing logistics support to the engineers, it will also provide organizational maintenance support to them. This causes no particular problem for common equipment. No engineer equipment repairers exist in the maneuver unit's maintenance sections lower than the brigade level except those at the engineer battalion. It may be unrealistic to expect the maneuver unit to provide all logistics support. The platoon may have to coordinate for maintenance assistance from its parent unit. When the engineers are providing logistics support, they will also provide organizational maintenance support.

Doctrinally, recovery is the responsibility of the owning unit. However, it is virtually impossible for engineers working in the TF area to recover their own vehicles. In the defense, the TF will have to provide recovery support at least back to the unit maintenance collection point (UMCP) located near the combat trains. If repairs cannot be made at the UMCP and the time and situation permit, the engineer's maintenance-support team (MST) can come forward and recover the down vehicle. The TF could provide additional recovery back to the BSA for repairs. The engineer battalion maintenance officer (BMO) or technician must ensure that each BSA has direct-support (DS) mechanics capable of repairing engineer equipment.

In the offense, the TF recovers the vehicle to the main supply route (MSR). Depending on the situation, the TF either turns vehicle recovery over to the engineers or recovery assets take the vehicle to the BSA. It must be clear in the TF SOP or order which MST has recovery responsibility for the engineer vehicles. The priority the TF commander places on the engineer assets determines the priority they are recovered in.

Engineer companies should have maintenance contact teams that are familiar with engineer heavy equipment to routinely visit work sites. The communications repairman and armorer can accompany this contact team to help anticipate problems and provide support before the mission is jeopardized.

## MEDICAL SUPPORT

Medical support and evacuation must be coordinated for each operation and normally come from the nearest company/team (see *Table 6-2, page 6-8)*. The platoon's aidman or combat lifesavers should be required to coordinate with the TF's medics and ensure that the medical requirements of the platoon are planned for and met. In the event of a mass casualty incident, the platoon may have to designate a platoon vehicle as a litter vehicle.

Table 6-2. Command and support relationships and logistics support

| Logistics Support | Class III | Class IV/V | Class VIII Medical | Maintenance | Emergency Resupply |
|---|---|---|---|---|---|
| Parent unit (OPCON, DS, GS) | Company LOGPAC | Basic loads are resupplied from the LOGPAC.<br><br>Mission loads are supplied by the supported unit. | Parent unit | Parent unit performs maintenance on all equipment. | Parent unit |
| Company/team (attached) | Company/team LOGPAC | Basic loads are resupplied from the LOGPAC.<br><br>Mission loads are supplied by the supported unit. | Company/team | Company/team MST performs maintenance on common equipment.<br><br>Engineer MST performs maintenance on engineer-specific equipment. | Company/team |
| TF (attached) | TF support platoon | Basic loads are resupplied from the LOGPAC.<br><br>Mission loads are supplied by the supported unit. | TF medical platoon | TF maintenance platoon performs maintenance on common equipment.<br><br>Engineer MST performs maintenance on engineer-specific equipment. | TF support platoon |

The platoon leader must—

- Understand the TF's evacuation and medical support plans.
- Understand the engineer company's evacuation and medical support plans.
- Know the locations of the—
  - Battalion aid station.
  - Jump aid station.
  - Personnel transfer points.
  - Casualty collection points.

# CHAPTER 7
# Command and Control

The platoon leader uses the $C^2$ processes to ensure that his platoon accomplishes its missions. This chapter provides the TTP needed to command and control the platoon and make sound tactical decisions.

## COMMAND AND SUPPORT RELATIONSHIPS

Command relationships can be operational control (OPCON) or attached. Support relationships can be DS or general support (GS). Both supporting and support parties must clearly understand what each of these mean (see *Table 6-2, page 6-8*). The following should be considered when using command and support relationships:

- When OPCON is used, the maneuver element—
  - Commands and controls the platoon.
  - Has the authority to place squads OPCON to companies/teams even though it is not considered advisable to do so with a mechanized platoon. Additionally, the TF cannot attach them to the company.
- When attached is used, the maneuver element—
  - Commands and controls the platoon except for limitations imposed by the attachment order.
  - Has responsibility for the platoon's logistics support except for personnel transfer and promotion.
- When DS is used—
  - The platoon receives its missions from the maneuver element.
  - The engineer company and the engineer battalion support and command the platoon.
  - The platoon works on the tasks that the maneuver element assigns.
  - The TF does not have the authority to split the platoon up and attach squads to companies/teams.
  - Logistics support for the platoon comes from engineer higher HQ.

- When GS is used, subordinate commanders must request support from the engineer commander on a task-by-task basis, and the senior commander sets priorities and assigns tasks to the GS unit. GS is seldom used at TF level. When an engineer commander wants to retain control over the engineers, they are retained in GS of maneuver forces.

The TF always provides the materials necessary to support engineer operations in its sector for offensive and defensive missions. This is a key logistics issue that must be understood at the start of any $C^2$ discussion. Even though the engineers determine requirements and use the material, supplying it remains a TF responsibility regardless of the command or support relationship (see *FM 5-100*).

## COMMUNICATIONS

Engineers must operate on the following frequency modulated networks:

- TF.
- Platoon internal.
- Engineer company.
- Maneuver company/team.

Engineers have a two-net capability to support their communication requirements. In the DS or GS relationship, the engineer platoon should operate the first network on the engineer command network. The second network, the platoon's internal network, should be used to control the platoon's assets. In the attached or OPCON relationship, the engineer platoon should operate the first network on the engineer command network or supported maneuver company/team command network based on the task organization. The platoon leader uses his platoon's internal network to enter other networks for coordination and situation reporting. Before the platoon leader leaves his platoon's internal network, he must first hand over control of the network to the platoon sergeant to control the platoon's assets.

The platoon submits the following reports:

- Routine reports:
  - Commander's situation report (SITREP)—identifies changes to a unit's tactical situation.
  - Sensitive-items report—gives the results of a serial-number check on all sensitive items.
  - Supply-status report—informs the commander and staff of a subordinate unit's supply status.
  - Personnel-status report—is used to maintain the accountability and status of assigned and attached personnel and generate requests for replacements.

  - Maintenance-status report—provides information on the operational status of equipment and is used to generate requests for maintenance and parts.
- Event-driven reports:
  - Spot report (SPOTREP)—identifies any known or suspected enemy activity.
  - Reconnaissance report—provides technical and tactical engineer data on a point/area target.
  - Countermobility report—gives the status of all obstacles.
  - Mobility report—gives the status and conditions of lanes, routes, bypasses, crossing sites, air lines of communication, as required, and ground lines of communication. It also includes status changes and the upgrade/downgrade of previously reported trafficability of locations/sites.
  - Survivability report—provides the detailed status of the unit's survivability level (for example, the number, type, and protective level (hull or turret) of positions).
  - Class IV/Class V special report—is used to keep track of critical engineer classes of supplies. For example, a platoon establishes a mine dump and keeps track of the types and quantities of mines available and the amounts issued.
  - Flash SITREP—is used to notify the next higher HQ when a subordinate unit becomes decisively engaged or has critical shortages that prevent mission accomplishment.
  - NBC report—is used to report NBC events.
  - Shelling report (SHELREP)—provides information on enemy indirect fire.
  - Enemy prisoner of war (EPW) report—is used to account for prisoners captured.
  - SCATMINE report—is used to notify affected units of the emplacement of SCATMINEs.
  - Meaconing, intrusion, jamming, and interference (MIJI) report—is used to notify the Intelligence Officer (US Army) (S2) when the reception of radio signals is hindered, confused, or distorted by any external source or when instructions are received from a station that cannot authenticate.
  - Splash report—is used to report downed friendly aircraft.
  - Casualty report—identifies battle and nonbattle casualties to the parent unit and alerts medical-treatment facilities of incoming casualties.
  - Casualty-feeder report and witness-statement report—provide detailed eyewitness information on a casualty. They are used to account for casualties and provide administrative information for use in notifications and awards.

## ENGINEER INTEGRATION

The company commander of a mechanized engineer company normally associated with a battalion/TF is also the staff engineer and advisor to the TF commander. Likewise, in a light TF, the staff engineer advisor to the TF commander is the light engineer platoon leader. The staff engineer integrates engineers into the TF's planning process. He coordinates with other BOS to ensure engineer synchronization. During the preparation of the battle, the staff engineer ensures that soldiers and equipment are in the correct location and on time for linkups and rehearsals. He then executes the engineer mission to support the operation. The platoon is the lowest level engineer unit that can still effectively accomplish independent tasks. The platoon usually operates under the control of an engineer company; however, it can be attached to a maneuver company/team.

The platoon leader must keep both the engineer company and maneuver company/team commanders informed on critical tasks. He must continuously forward and update tactical and technical information upward and laterally by all available means.

In heavy forces, the TF normally receives an engineer company in support, and the company commander is the TF engineer. In light forces and in some special situations in heavy forces, the TF only receives an engineer platoon, and the platoon leader is the TF engineer. When the platoon leader is acting as the TF engineer, he should consider the following information:

- Orders should include when and where the TOC is to be established. If the TOC is to be moved before the platoon arrives, this should also be included.
- Orders should identify an assembly area where the platoon can be consolidated.
- The entire platoon must **not** go to the TOC.

## PLATOON LEADER'S RESPONSIBILITIES

Once the platoon leader arrives at the TOC, he should—

- Check in with the S3 or company commander and report the platoon's status and location and any problems.
- Provide the status of equipment and the number of soldiers to the operations noncommissioned officer (NCO).
- Obtain the TF's and the team's signal operation instructions (SOI) and enter the maneuver force's network.

- Ensure that the communications equipment is set for the correct variable for secure operations and is compatible with the supported unit's communications equipment.
- Check with the operations sergeant for any changes to routine or required reports and SOP items.
- Be familiar with the TF SOP. **This is not the time to start studying the SOP.**
- Provide the status of obstacles being constructed.
- Check on format requirements for the periodic report. He should also check to see that TOC personnel are familiar with the reports being sent and know how to post them to their charts or overlays.
- Determine if a specific individual at the TOC has been designated as an expediter or overwatcher for engineers in the sector requiring assistance.
- Provide any terrain, route, or other information acquired from the reconnaissance unit or personnel enroute to the S2 or company/team commander.
- Obtain such information as the current enemy situation, weather and light data, and terrain details, to include topographic products, from the S2 or company/team commander.
- Provide any priority intelligence requirements (PIR) or information requirements (IR) needed for upcoming operations to the S2 or company/team commander.
- Stop at the combat trains and coordinate for classes of supplies and services when joining the maneuver force.
- Obtain frequencies and call signs for the administrative and logistics net control stations (NCSs).
- Provide a platoon roster and identify any critical shortages to the Adjutant (US Army) (S1) or operations sergeant.
- Review the TF SOP for reporting and evacuating casualties.
- Coordinate with the S4 or company/team 1SG and review the platoon's logistics status. The channel used for operational logistics for the platoon varies based on the command and support relationships.

**NOTE: The platoon leader should remember that if a problem arises which will prevent the accomplishment of the mission or if normal channels are not working fast enough and logistics become an operational issue, the S3 or commander should be apprised of the problem.**

As a minimum, the platoon leader must be present when the—

- Maneuver force receives the WO.
- Commander conducts reconnaissance.
- Commander makes and announces a decision and explains the intent.
- OPORD is received.

## MANEUVER FORCE'S CONTROL

Normally, the platoon is kept under the engineer company's control and works for the battalion/TF commander. A major consideration in planning for the use of engineers is to maintain platoon integrity, keeping the entire platoon under the platoon leader's control. When maneuver forces are moving on two axes during mechanized offensive operations, the platoon will not place a part of the platoon on each axis. During the planning process, the engineer planner must weigh the main effort and determine which axis is critical to the mission and requires engineer assets. Placing an engineer NCO with the scout platoon when it is performing reconnaissance can help provide more specific information on the obstacles that the scouts find. Obstacle breaching can thus be preplanned. See *Appendix A* for basic formations, movement techniques, and hand-and-arm signals and *Appendix B* for checklists.

## ENGINEER ATTACHMENTS

If the platoon has been augmented with AVLBs, they normally come in pairs. This is for both maintenance and tactical reasons. The AVLBs should be placed with the element that needs a gap-crossing capability. In the deliberate attack, AVLBs can be integrated individually into attack formations. Employment as a pair allows the squad leader to control them. If the AVLBs are not immediately required, they should be returned to the company's control. Curbs on the AVLB must be removed if M1 tanks with rollers are to cross the AVLB, and the roller's dog bone must be raised. If a CEV is attached to the platoon, it normally moves with the platoon.

Heavy equipment and wheeled vehicles should move with the platoon only on tactical road marches over good roads. During a movement to contact or over rougher terrain, METT-T determines where and what heavy equipment moves in the formation.

# APPENDIX A

# Basic Formations, Movement Techniques, and Hand-and-Arm Signals

The platoon uses a variety of mounted and dismounted formations and movement techniques to maneuver on the battlefield. This appendix gives examples of many of the basic formations, movement techniques, and hand-and-arm signals the platoon leader could use. It is not designed to be all encompassing. For more information on these topics, see *FMs 7-8, 5-34, and 71-1.*

## MOUNTED MOVEMENT TECHNIQUES

The mounted engineer platoon must be proficient in moving with its maneuver counterpart. In the following paragraphs, formations, movement techniques, and actions taken during movement for the mounted engineer platoon are discussed.

### *WEDGE*

The engineer platoon almost always maneuvers as part of another larger formation. Normally, this will either be the parent engineer company or a maneuver company/team. *Figure A-1, page A-2,* shows the formation that the platoon is most likely to use. The platoon wedge provides the most defensible formation with the easiest $C^2$. Ordinarily, the platoon follows a maneuver platoon if task-organized to a maneuver company/team. The platoon might lead if it is part of the engineer company formation. In either case, the wedge is the best formation to use if enemy contact is likely.

The platoon frequently receives assets from the assault and obstacle (A&O) platoon. These could include the AVLB, CEV, or armored vehicle-launched mine-clearing line charge (AVLM). These additional vehicles can strain the $C^2$ capability of the platoon. *Figure A-2, page A-2,* shows the technique of "tethering". The AVLB, CEV, and AVLM are tethered to individual engineer squads. Each squad leader has the responsibility of commanding and controlling an A&O platoon asset. This improves the $C^2$ of the platoon and provides each A&O asset with a security element as it moves across the battlefield.

*Figure A-3, page A-3,* shows a simple wedge formation. Note where the key leaders in the platoon are. The wedge formation uses the wingman concept. The platoon sergeant and leader are both in APCs and each has a wingman. This technique

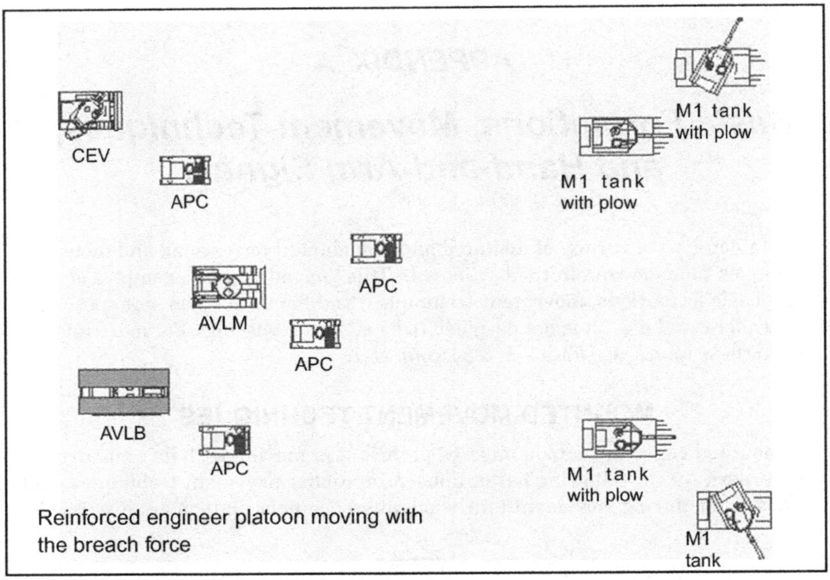


**Figure A-1. Engineer platoon with the breach force**

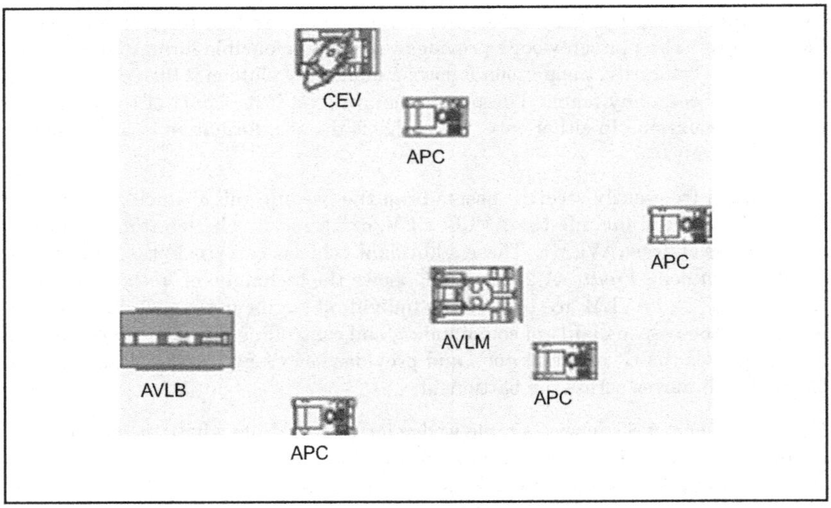


**Figure A-2. Modified mounted wedge formation**

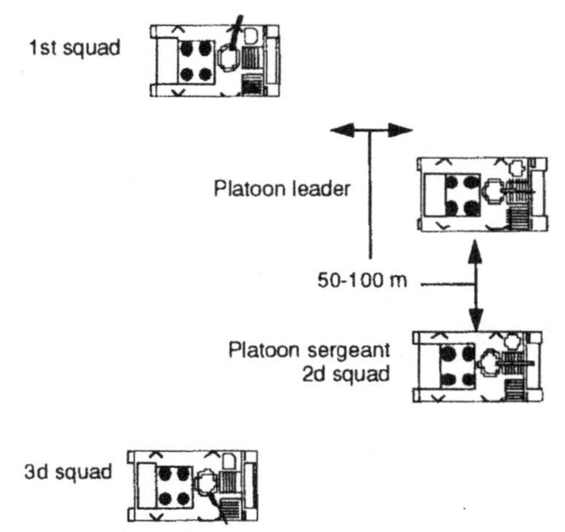


**Figure A-3. Mounted wedge formation**

also simplifies $C^2$ and movement. It has the added benefit of placing the most experienced leader in the platoon, the platoon sergeant, forward with the platoon. It also places the platoon sergeant where he can easily take over in the absence of the platoon leader.

### COLUMN

*Figure A-4, page A-4,* depicts a column formation. This formation is used when enemy contact in not expected. This formation maximizes $C^2$ and the speed of the formation. Normally, the platoon transitions from the column to the wedge as enemy contact becomes more likely.

### LINE

*Figure A-5, page A-4,* shows the platoon in a line formation. This formation is designed to maximize the forward firepower of the platoon. The platoon transitions from the wedge to the line as enemy contact becomes eminent. Engineer platoons do not generally use this formation. However, if the platoon is tasked to suppress a dismounted threat while another platoon maneuvers, this formation works well.

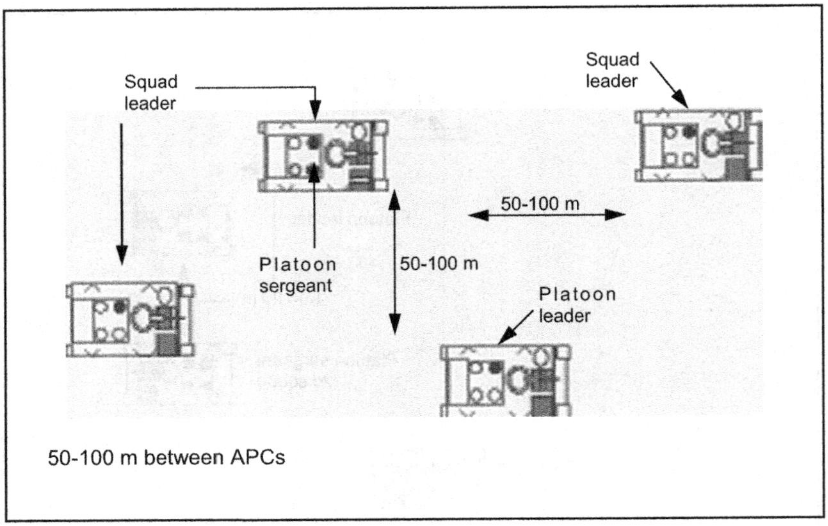


**Figure A-4. Mounted column formation**

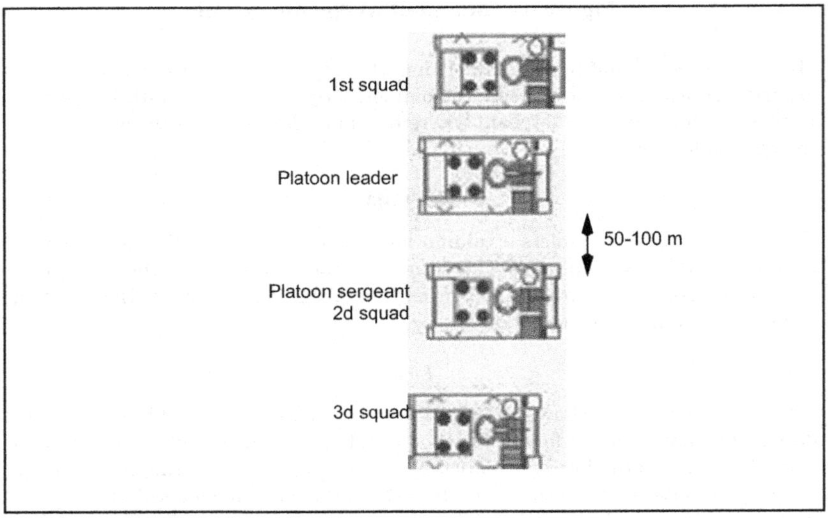


**Figure A-5. Mounted line formation**

### *ECHELON*

The echelon formation is used when the platoon is on the flank of the company and the enemy threat is also from the flank. *Figure A-6* shows an echelon right formation. The echelon formation can also be used on the left flank of the company and, in that case, it would be the mirror image of *Figure A-6*. This formation maximizes the platoon's firepower to the flank.

### *V*

The V-formation is a variation of the wedge. It is used when there is a significant threat of command-detonated mines or explosives. The formation shown in *Figure A-7, page A-6,* would be used during a route-clearance operation. It allows the platoon to secure the flanks of the road while two squads clear the route. This formation also identifies command-detonated-mine firing wires or ambushes before the enemy can attack the element on the route. It is not generally used when there is a significant enemy threat.

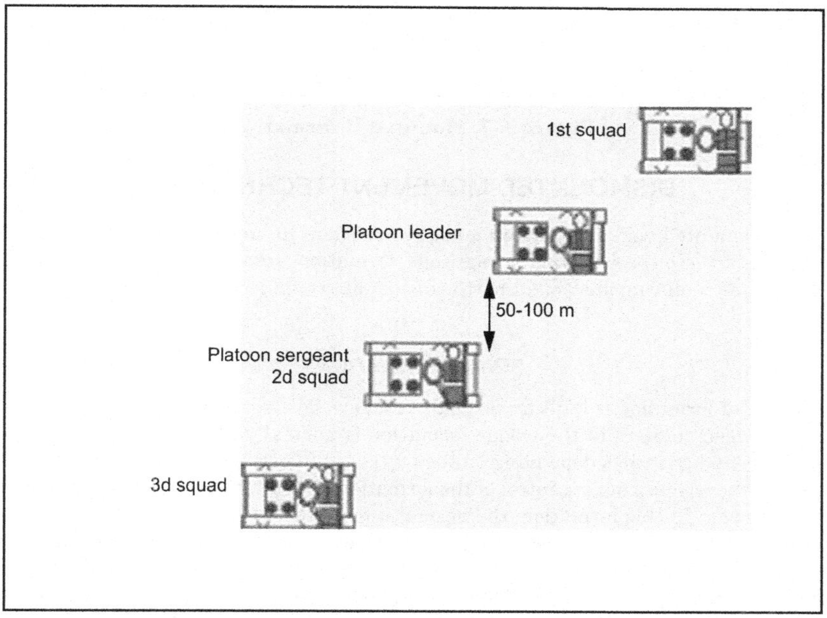


**Figure A-6. Mounted echelon formation**

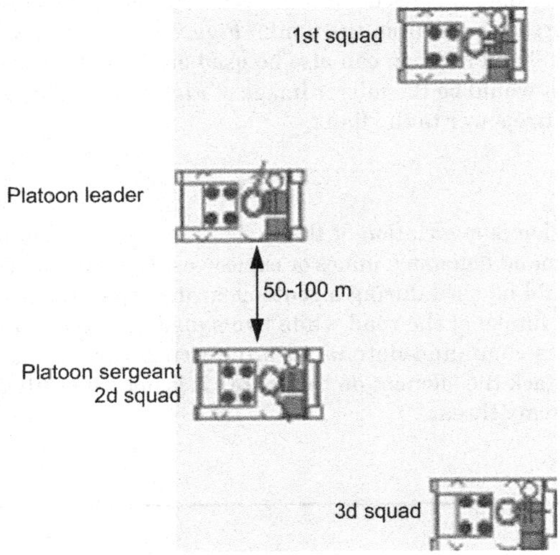


**Figure A-7. Mounted V-formation**

## DISMOUNTED MOVEMENT TECHNIQUES

The dismounted engineer platoon must be proficient in moving with its maneuver counterpart. In the following paragraphs, formations, movement techniques, and actions taken during movement for the dismounted engineer squad and platoon are discussed.

### *SQUAD FORMATIONS*

The squad formation is built from the fire-team wedge (see *Figure A-8*). The interval between soldiers in the wedge formation is normally 10 meters. The wedge expands and contracts depending on the terrain. When rough terrain, poor visibility, or other factors make control of the formation difficult, the squad uses the modified wedge. In this formation, the normal interval between soldiers is reduced so that all team members can still see their team leader and each team leader can see their squad leader. The sides of the wedge can contract to the point where the wedge resembles a single file. When moving in less rugged terrain, where control is easier, soldiers expand the formation or resume their original positions.

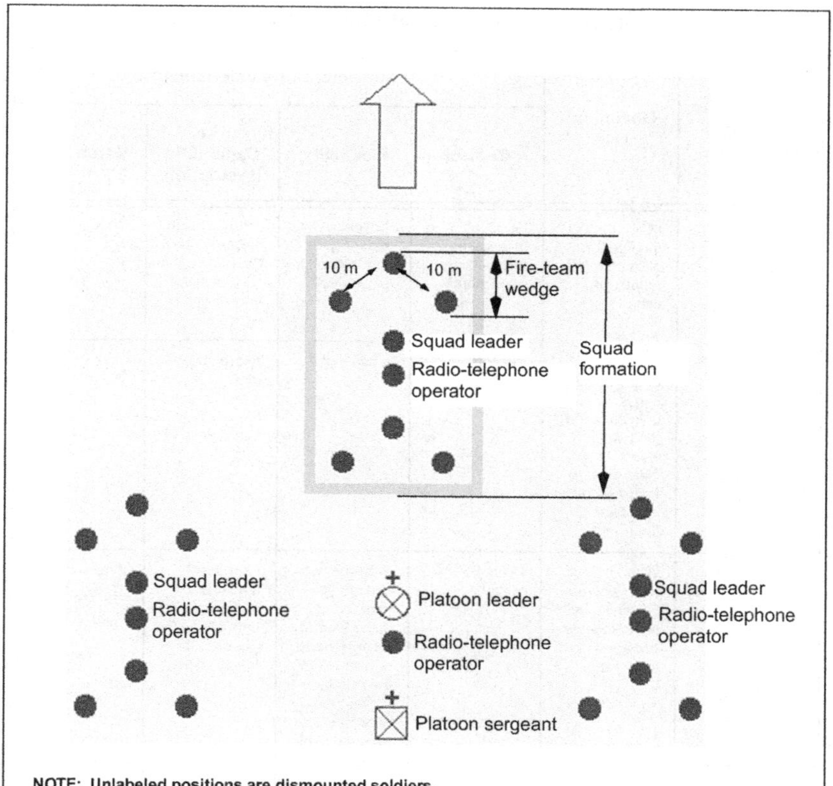


**Figure A-8. Dismounted-platoon wedge formation**

### *PLATOON FORMATIONS*

Platoon formations are built from multiple squad formations. These formations include the platoon column, line, V, and wedge. The platoon leader selects the best formation based on his METT-T analysis. *Table A-1, page A-8*, shows a comparison of each formation and its use.

**Table A-1. Comparison of platoon formations**

| Dismounted Movement Formations | Movement Use | Movement Characteristics | | | |
|---|---|---|---|---|---|
| | | Control | Flexibility | Fire Capability/ Restriction | March Rate |
| Column | When the platoon performs primary movement formations | Good for maneuver (fire and movement) | Provides good dispersion laterally and in depth | Allows limited firepower to the front and rear, high volume to the flank | Good |
| Line | When the enemy situation is unknown and the leader wants all soldiers forward for maximum firepower to the front | Difficult | Is minimal | Allows maximum firepower to the front, little to the flanks and rear | Slow |
| V | When the enemy situation is vague, but contact is expected from the front | Difficult | Provides two squads up front for immediate firepower and one squad to the rear for movement upon contact from the flank | Allows an immediate heavy volume of firepower to the front or flanks | Slow |
| Wedge | When the enemy situation is vague, but contact is not expected | Difficult but better than the platoon V and platoon line | Enables the leader to make contact with the smallest element and still have two squads to maneuver | Provides a heavy volume of firepower to the front or flanks | Slow but faster than the platoon V |
| File | When visibility is poor due to terrain or light | Easiest | Is the most difficult formation from which to maneuver | Allows immediate fires to the flanks; prevents focused fires to the front and rear | Fastest |

### Wedge

The dismounted wedge is used when the enemy situation is vague and enemy contact is likely (see *Figure A-8, page A-7)*. This formation allows a large volume of fire around the formation. Generally, at least one squad or element will be free to maneuver from the wedge after contact.

### Column

The column formation is the platoon's primary movement formation (see *Figure A-9, page A-10*). It provides good dispersion both laterally and in depth and simplifies control. The lead squad is the base squad for fire control.

### Line

The line formation allows the delivery of maximum fire to the front but little fire to the flanks (see *Figure A-10, page A-11)*. This formation is hard to control and does not lend itself well to rapid movement. It is the basic platoon assault formation during an attack.

### V

The V-formation has two squads up front to provide a heavy volume of fire on contact (see *Figure A-11, page A-12)*. It also has one squad in the rear that can either overwatch or trail the other squads. This formation is hard to control, and movement is slow.

### File

The file formation gives maximum control to leaders and is used for speed during movement (see *Figure A-12, page A-13)*.

### Traveling

The dismounted engineer platoon supporting a maneuver TF normally travels as part of one of the larger maneuver elements. When breaching assets are needed forward with the lead company, the engineer platoon travels behind the lead platoon for security (see *Figure A-13, page A-14)*. If each maneuver company needs a breaching capability, the dismounted engineer platoon can be broken into squad-sized elements, where each squad travels with a maneuver company. In this case, the platoon HQ travels with either the battalion tactical operations center or the main effort for $C^2$.

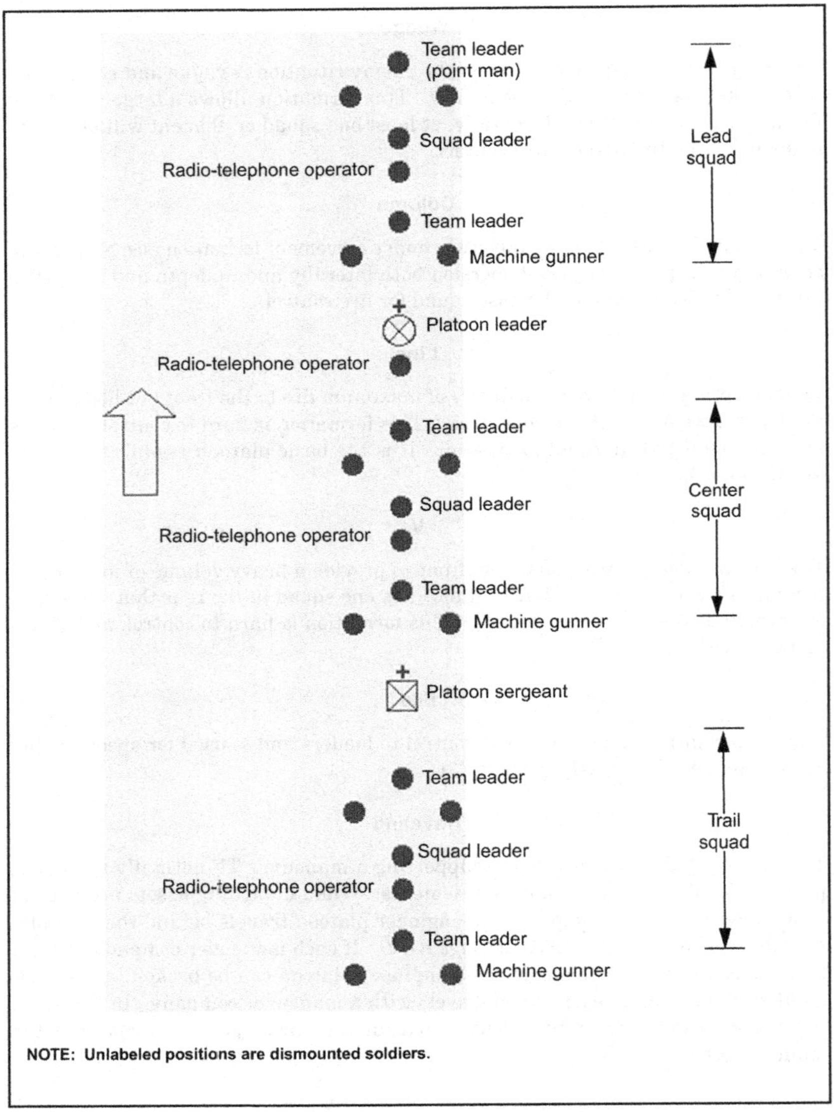


**Figure A-9. Dismounted column formation**

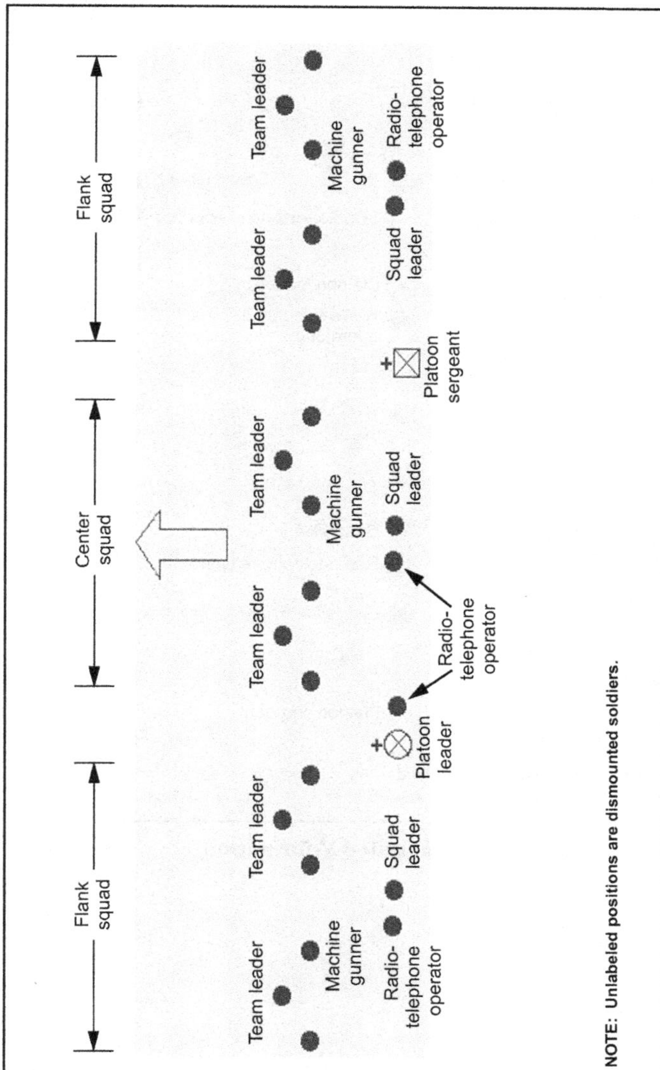


Figure A-10. Dismounted line formation

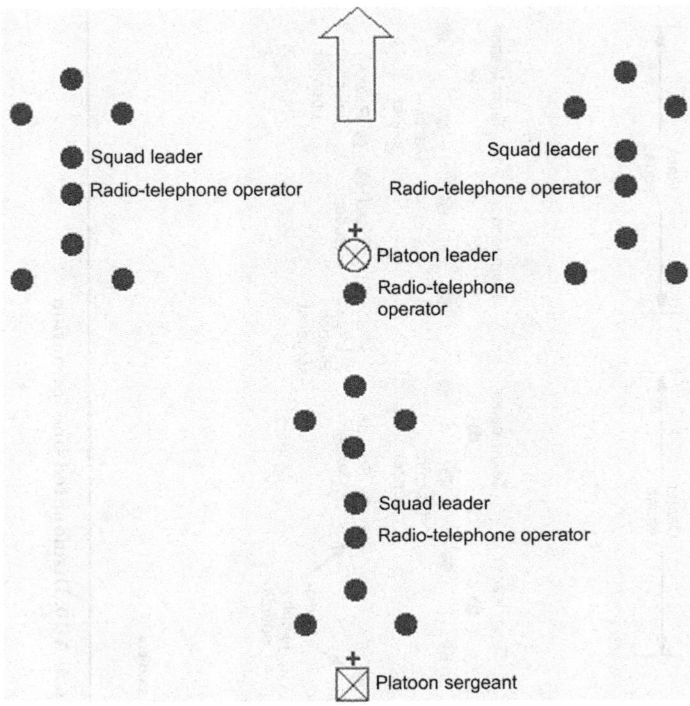


**NOTE: Unlabeled positions are dismounted soldiers.**

**Figure A-11. Dismounted V-formation**

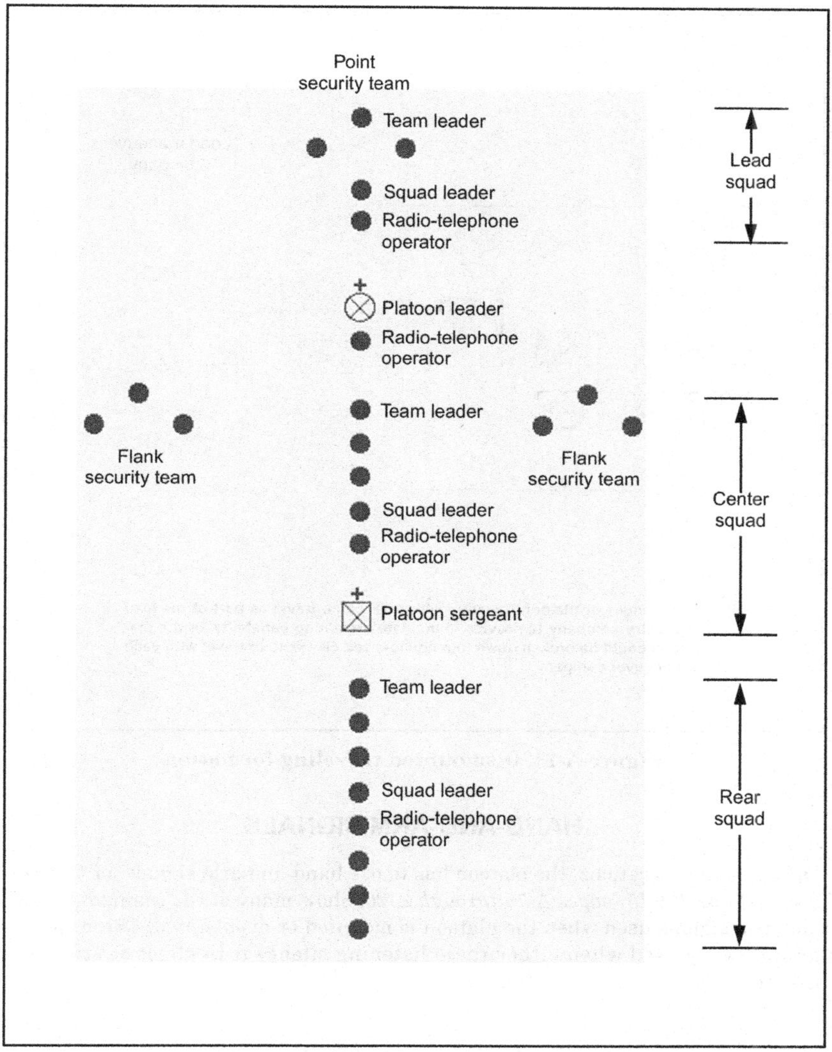


**Figure A-12. Dismounted file formation**

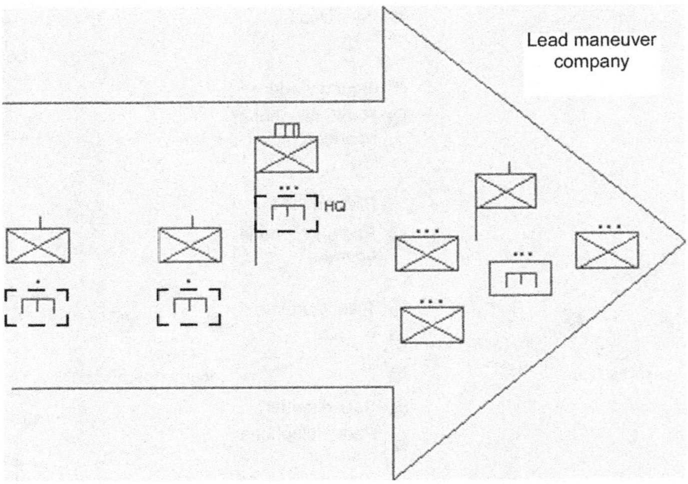


**NOTE:** The engineer platoon traveling with a TF could travel as part of the lead infantry company to provide immediate breaching capability, or the platoon could be broken down into squad-sized elements to travel with each maneuver company.

**Figure A-13. Dismounted traveling formation**

## HAND-AND-ARM SIGNALS

During many operations, the platoon has to use hand-and-arm signals for $C^2$. *Figures A-14 and A-15, pages A-15 through A-20,* show many of the command hand-and-arm signals used when the platoon is mounted or dismounted. These signals are normally used when either radio listening silence is in effect or stealth is needed.

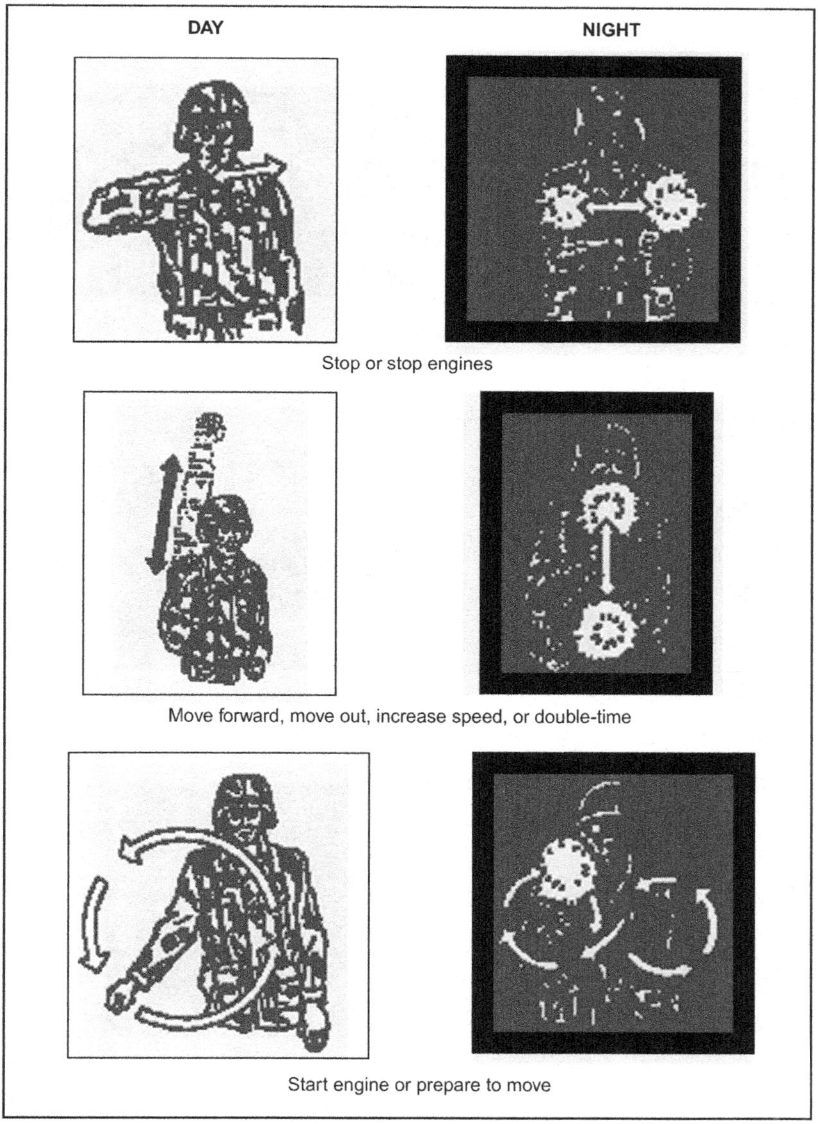


**Figure A-14. Hand-and-arm signals**

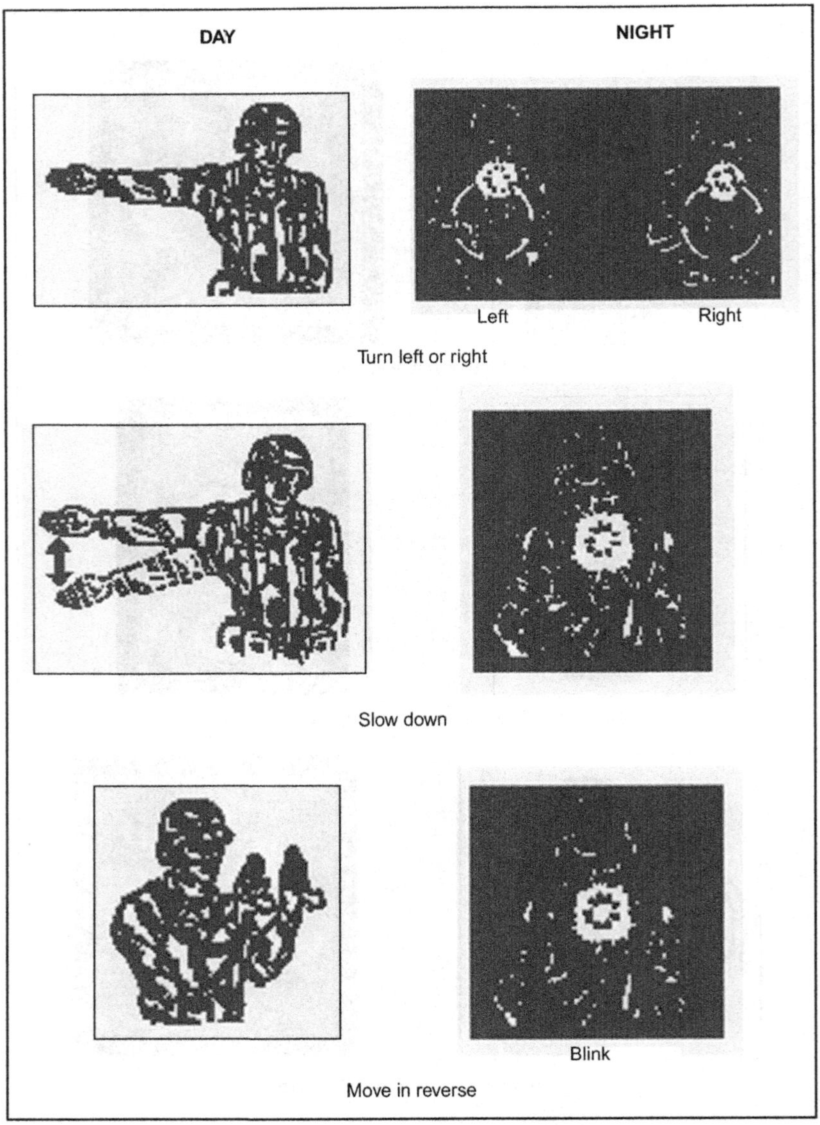


**Figure A-14. Hand-and-arm signals (continued)**

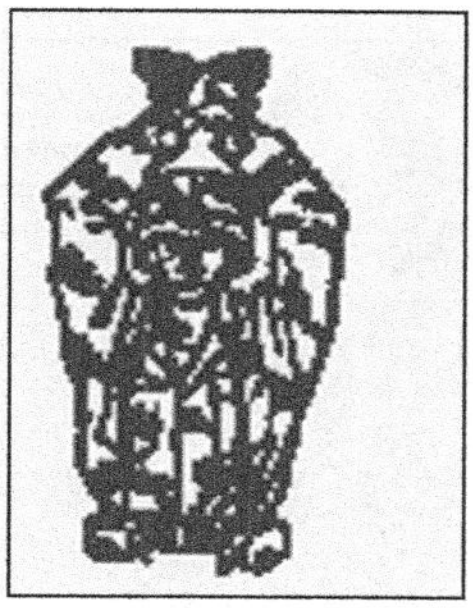

Disregard previous command

Attention

I do not understand

I am ready

Close the distance between vehicles and stop

**Figure A-14. Hand-and-arm signals (continued)**

Stop

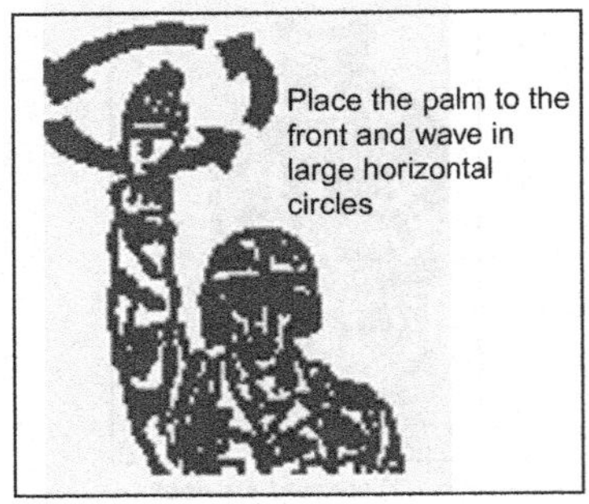


Assemble

Enemy is in sight

Commence firing

Cover our move

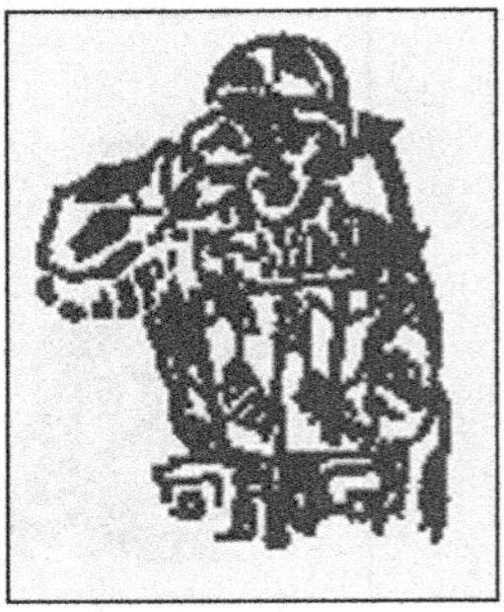

Cease firing

**Figure A-14. Hand-and-arm signals (continued)**

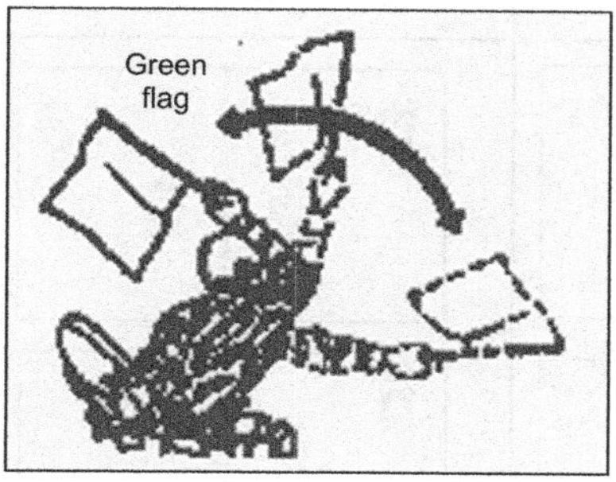


Move out

Enemy is in sight

Form a line

**Figure A-14. Hand-and-arm signals (continued)**

| Mounted Movement Formations | Hand-and-Arm Signals | Formation Examples |
|---|---|---|
| V |  | 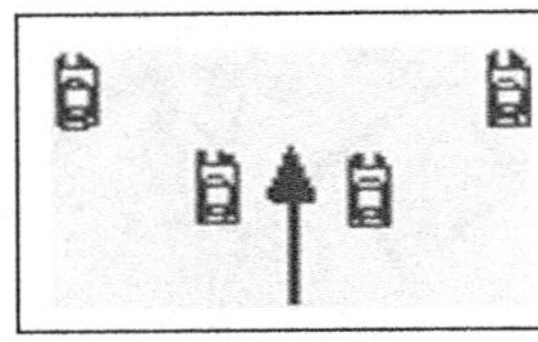 |
| Echelon right (left) |  | 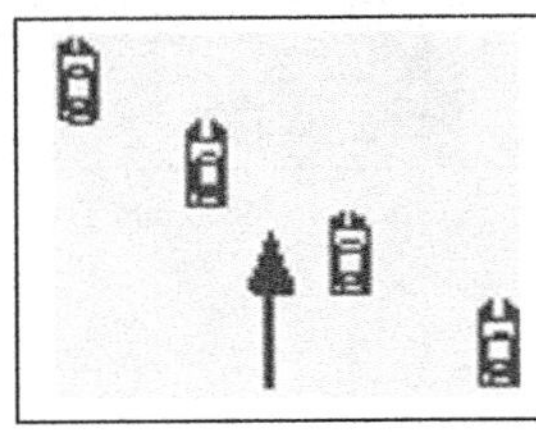 |
| Wedge |  | 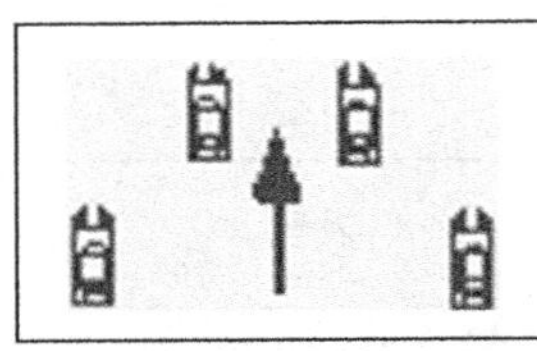 |
| Line |  | 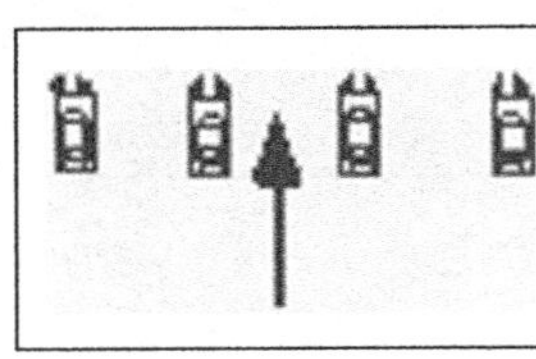 |
| Column |  | 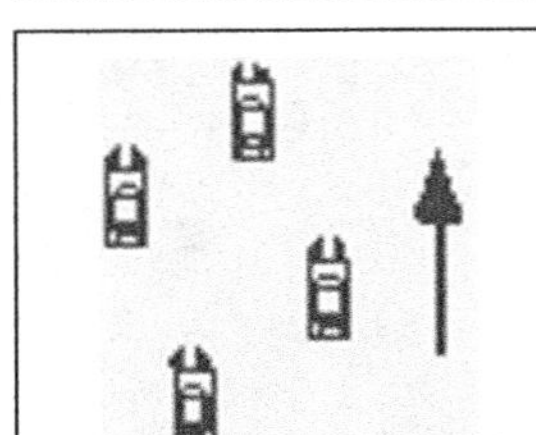 |

**Figure A-15. Mounted movement formations and hand-and-arm signals**

# *APPENDIX B*

# *Checklists*

This appendix gives several checklists that the platoon and its leaders will find useful. These checklists are not all encompassing and should be used to augment local SOPs, not replace them. Where units do not have standardized checklists, these may be helpful in developing local unit SOPs.

## LOCAL SECURITY AND JOB-SITE SECURITY CHECKLISTS

When the platoon halts, the platoon leader establishes and maintains local security. He assigns each squad a sector of the perimeter to ensure mutual support and to cover all gaps by observation and fire. The platoon leader designates OPs and squad leaders designate OP personnel. OPs communicate with the platoon's command post and warn the platoon of the approaching enemy before it is attacked. Frequently, the platoon will be in a tactical halt while waiting for an operation to start.

When the platoon is halted, the platoon leader establishes a priority of work, to include—

- Positioning crew-served weapons, protective obstacles, and chemical alarms.
- Constructing individual and crew-served fighting positions, if the halt is for an extended period.
- Setting up wire communications between the squads and the platoon's command post. Radio silence should be maintained at the halts for OPSEC.
- Coordinating with adjacent units.
- Preparing range cards.
- Clearing fields of fire.
- Distributing ammunition, rations, water, supplies, and special equipment.
- Conducting PMCS on weapons, vehicles, and equipment.
- Inspecting the platoon's members and equipment.

- Rehearsing critical aspects for the upcoming mission.
- Test firing small-arms weapons, if the tactical situation permits.
- Conducting personal hygiene and field sanitation.
- Instituting a rest plan.
- Completing the work priorities as time permits.

Job-site-security actions are similar to those actions that are taken at a halt. However, job-site security is specifically designed to ensure that the platoon is able to work without being harassed.

Typically, the platoon will use this checklist when constructing obstacles and field fortifications. Before moving to the job site, the platoon leader informs everyone of the warning signals, code words, and pyrotechnics signals to be used. Upon arrival at the job site, the platoon—

- Occupies a job-site overwatch position.
- Dispatches reconnaissance/minesweeping/NBC teams to clear the job site.
- Establishes local security.
- Establishes escape routes; identifies possible enemy avenues of approach; and places OPs, chemical alarms, and crew-served- and antitank-weapons positions to cover enemy avenues of approach.
- Divides the job site into defensible sectors and assigns them to the squads.
- Establishes communications with the engineer company.
- Starts work.

## PRECOMBAT CHECKLISTS

Each individual should have—

- Load-bearing equipment.
- A chemical protective mask with an M256 chemical-detection kit, an M13 decontamination kit, a waterproof bag for the mask, M8 chemical-detection paper, and three chemical-antidote injectors.
- A kevlar helmet with a cover and a band.
- An individual weapon with magazines and the zero posted in the pistol grip and a cleaning kit.
- A ruck sack and duffle bags packed with military clothing and table of allowance (TA)-50 equipment.

- Wire gauntlets.
- A fragmentation vest.

Each leader should have—

- A map and a compass.
- Signal operating instructions.
- His unit's SOP.
- *FMs 5-34 and 5-10* in their uniform cargo pockets.
- *FMs 20-32, 90-13-1, and 90-7* in their ruck sack.
- Other FMs, as required or needed.
- A watch and a solar-powered calculator.
- A notebook, pens, and pencils.
- Binoculars.
- Night-vision goggles.
- Casualty cards.
- A laminated range card.
- *DA Forms 1355 and 1355-1-R.*

All vehicles should have—

- Material, tied down and loaded according to the load plan.
- Fuel tanks, full.
- Preventive maintenance checks and services (PMCS), completed and dispatched.
- Packaged POL products, stowed.
- Water and extra fuel cans, full.
- Meals, ready-to-eat (MREs), stored.
- Basic-issue items (BIIs) and additional authorized list (ALL) items, inventoried, cleaned, and stowed.
- Spare parts and tires or track blocks, stowed.
- First-aid kits, inspected and stowed.
- A vehicle log book, the vehicle's TM, and *DA Forms 2404,* stowed.

- A camouflage screen system, stowed.
- Chemical-detection equipment, mounted.
- A tow cable with locking pins, stowed.
- VS17 marking panels for emergency marking.
- Ammunition evenly distributed and demolitions properly segregated and prepared for use.

## UNIFORM-, DUFFLE-BAG-, AND RUCK-SACK-PACKING CHECKLISTS

A field uniform should include—

- A kevlar helmet with a cover and a band.
- A fragmentation vest.
- A pistol belt with ammunition pouches.
- A canteen with a cover and a cup, on the right hip.
- Ear plugs.
- A flashlight, on the right suspender strap (optional).
- A holster with a lanyard (if armed with a pistol), on the right side of the pistol belt.
- A bayonet, on the left side of the pistol belt.
- A compass, on the left side of the pistol belt.
- An M17A1 protective mask, on the left hip, and an M25 mask under the left arm.
- A battle dress uniform (BDU) with black boots.
- A field jacket.
- A chemical-decontamination kit connected to the protective mask.
- A nerve-agent antidote.
- A first-aid pouch, on the left suspender strap.

One duffle bag should contain—

- Three sets of BDUs.
- Three pairs of underwear.

- Three pairs of wool socks.
- Three T-shirts.
- One towel.
- Two laundry bags.
- One waterproof bag.
- One pair of boots with shoe polish, a brush, and a sewing kit.

The additional duffle bag should contain—

- Five tent pegs (10 for officers).
- Three tent poles (6 for officers).
- One tent rope (2 for officers).
- One sleeping bag.
- One pair of trouser suspenders.
- Two pairs of coveralls.
- One field jacket with a liner.
- One shelter half (2 for officers).
- One wool blanket.
- One pair of field pants.
- One pair of field pants with a liner.
- One cold-weather mask.
- One pair of extreme cold-weather boots.

The ruck sack should contain—

- Wet-weather pants and a parka.
- Overshoes.
- A weapons cleaning kit.
- Rations.
- A waterproof bag.
- A mission-oriented protective posture (MOPP) suit with gloves.
- Cold-weather mittens, winter use only.
- A parka with a liner and a hood, winter use only.

- A wool sweater, winter use only.
- A pile cap, winter use only.
- A scarf, winter use only.
- A towel and toilet articles.
- A pair of underwear.
- A T-shirt.
- Two pairs of wool socks.
- A BDU hat.
- An entrenching tool.

# APPENDIX C

# Breaching Case Studies

*FM 90-13-1* and *Chapter 3* provide doctrinal overviews for breaching, breaching tenets, and breaching theory. The following case studies are designed to provide the platoon with instructive examples of dismounted and mounted breaching. Either case study could be applied to many different field applications. For example, the dismounted case study could be used to examine covert breaching, assault breaching, or deliberate breaching by light engineers. Likewise, the mounted case study could apply to both deliberate and in-stride breaching. These case studies are not designed to be prescriptive. They are designed to illustrate the thought process that the platoon leader must go through to successfully breach.

All breaching operations follow a similar pattern. They differ in technique. During breaching, the—

- Platoon leader determines the obstacle location and its composition.
- Platoon leader selects the breach site and lane locations.
- Platoon leader orders the breach to start, if the platoon is the primary reduction force.
- Squad prepares to breach, if the platoon is the backup reduction force.
- Squad breaches the obstacle.
- Squad marks the lane.
- Platoon leader reports the lane location to the company commander.

## DISMOUNTED CASE STUDY

The company commander orders the dismounted platoon leader that is attached to him to breach an obstacle. The company is assaulting an enemy objective. The platoon leader moves to a location where he can see the obstacle and determines that it is a concertina fence and antitank minefield. He orders a squad forward to conduct reconnaissance and ascertain the best lane locations. The squad leader reports the locations and marks them with orange tape. The squad has determined that there are no antipersonnel mines in the minefield. The squad leader reports this to the platoon leader (see *Figure C-1, page C-2)*.

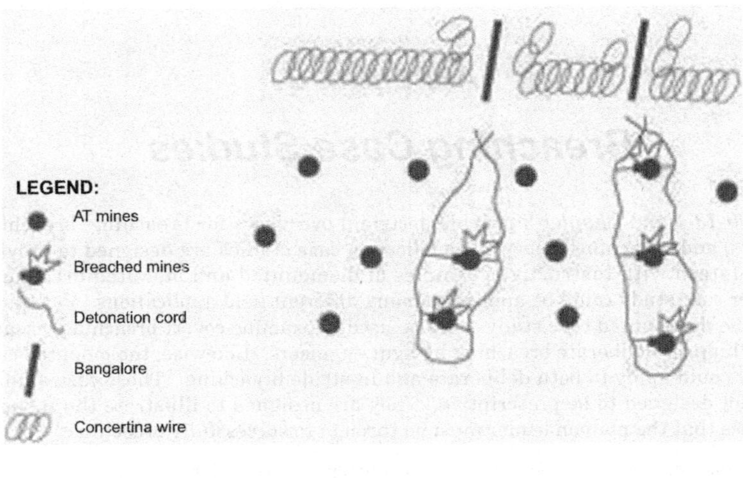


1 – The platoon leader determines the obstacle location and its composition.

2 – The platoon leader selects the breach site and lane locations.

3 – The squad breaches the concertina fence using cutters or Bangalores.

4 – The squad breaches the minefield using hand-emplaced explosives.

5 – The squad marks the lanes for the assault force.

6 – The platoon leader reports to the company commander that the lane is open.

**Figure C-1. Dismounted case study**

The platoon leader then orders the forward squad to remain at the obstacle as guides. He orders his second squad to come forward with two Bangalore sets. The first squad guides the second squad to the lane locations. Both squads breach the lanes in the concertina using the Bangalores and then move through the minefield placing charges to destroy antitank mines in the proposed lanes. The squads move to safety and initiate the explosives destroying the mines. Both squads provide security while the third squad comes forward to mark the lanes and guide the infantry company through the breach. The platoon leader keeps the third squad in reserve in case one of the squads fail or require additional help. However, in this scenario that was not necessary. The platoon leader observes the lanes and reports their completion to the infantry company commander who then starts his assault. The platoon leader then switches to the engineer company frequency and gives his company commander a situation report in reference to the lanes.

## MOUNTED CASE STUDY

The engineer platoon is fighting as part of an engineer company/team. The company/team, which includes the engineer line platoon, the A&O platoon, and a tank platoon equipped with track-width mine plows, is the TF breach force. The company commander decides to use the tank plows, initially for speed and security, to create the breach site and secure the far side of the obstacle. Then the engineer platoon widens and marks the lanes. The A&O platoon leader places the far-lane recognition signals and initially orients the TF on the breached lanes (see *Figure C-2).*

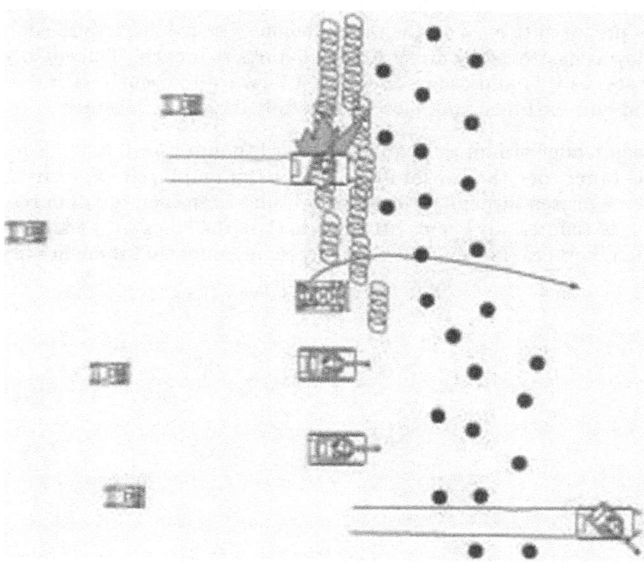

1 – The platoon leader determines the obstacle location and its composition.

2 – The platoon leader selects the breach site and lane locations.

3 – The tank crew breaches the obstacle with the tank plows.

4 – The platoon leader orders the AVLM forward to breach with the MICLIC when the enemy destroys a tank plow.

5 – The squad marks the lanes for the assault force.

6 – The platoon leader reports to the company commander that the lane is open.

**Figure C-2. Mounted case study**

In this scenario, the engineer company commander orders the engineer platoon forward to identify the obstacle location and its composition and potential lane locations. The platoon leader orders a dismounted element forward to conduct his reconnaissance. He reports his reconnaissance to the company commander. He states that the obstacle is a mixed minefield with antitank and antipersonnel mines, and a partial, multiple row of concertina fence is forward of the minefield. The platoon leader also reports two potential breach sites with an alternate site between them.

The engineer company commander orders the tank platoon to breach two lanes, based on the engineer platoon leaders' reconnaissance. He orders the engineer platoon to be prepared to back up the tank platoon. The northern tank equipped with a mine plow is destroyed by direct fire as it starts to breach. The engineer platoon leader observes this and orders his AVLM forward to breach with the MICLIC. The AVLM and the other tank plow successfully breach the obstacle.

The platoon leader orders two squads forward to mark each lane. These squads widen the lanes, per the initial plan, with hand-emplaced explosives after the assault force passes through. The platoon leader keeps one squad in reserve, with the AVLM, to counter any enemy attempts to close the lanes with artillery-delivered mines. He reports to the engineer company commander that the lanes are marked.

# Glossary

| | |
|---|---|
| & | and |
| + | plus |
| # | number |
| % | percent |
| 1SG | first sergeant |
| 1st | first |
| 2d | second |
| 3d | third |
| A&O | assault and obstacle |
| AA | axis of advance |
| AAR | after-action review |
| ABF | attack by fire |
| AC | hydrogen cyanide |
| ACE | armored combat earthmover, M9 |
| AD | air defense |
| ADA | air-defense artillery |
| ADAM | area denial artillery munition |
| AHD | antihandling device |
| ALL | additional authorized list |
| ammo | ammunition |

| | |
|---|---|
| AP | antipersonnel |
| APC | armored personnel carrier |
| app | appendix |
| arty | artillery |
| ASL | authorized stockage list |
| aslt | assault |
| AT | antitank |
| AT-3, AT-5, AT-8 | antitank missiles |
| ATD | antitank ditch |
| attn | attention |
| AVLB | armored vehicle-launched bridge |
| AVLM | armored vehicle-launched MICLIC |
| BDU | battle dress uniform |
| BII | basic-issue items |
| BMO | battalion maintenance officer |
| BMP | Soviet tracked amphibious infantry combat vehicle |
| BMP-1, BMP-2 | Soviet infantry fighting vehicles |
| BOS | Battlefield Operating Systems |
| BSA | brigade support area |
| BT | blade team |
| $C^2$ | command and control |
| CA | combined arms |
| CAM | chemical agent monitor |
| CATK | counterattack |
| CDR | commander |

| | |
|---|---|
| CEV | combat engineer vehicle |
| CG | phosgene |
| ℄ | centerline |
| COA | course of action |
| CP | checkpoint |
| CSS | combat service support |
| cu | cubic |
| DA | Department of the Army |
| decon | decontaminate/decontamination |
| DP | dispersing point |
| DS | direct support |
| DSA | division support area |
| EA | engagement area |
| EPW | enemy prisoner of war |
| FASCAM | family of scatterable mines |
| FEBA | forward edge of the battle area |
| FIST | fire-support team |
| FLOT | forward line of own troops |
| FM | field manual |
| FMFM | fleet Marine force manual |
| FS | fire support |
| FSO | fire-support officer |
| ft | foot (feet) |

| | |
|---|---|
| GA | tabun |
| GB | sarin |
| GD | soman |
| GEMSS | Ground-Emplaced Mine-Scattering System |
| GS | general support |
| HD | distilled mustard |
| HDP | hull-down position |
| HEMTT | heavy expanded mobility tactical truck |
| HL | mustard lewisite |
| HMMWV | high mobility multipurpose wheeled vehicle |
| HQ | headquarters |
| hr | hour(s) |
| IOE | irregular outer edge |
| IPB | intelligence preparation of the battlefield |
| IR | intelligence requirements |
| km | kilometer(s) |
| lb | pound(s) |
| LD | line of departure |
| LOGPAC | logistical package |
| LRA | local reproduction authorized |
| LRP | logistics release point |
| m | meter(s) |

| | |
|---|---|
| M-S | Miznay-Schardin |
| METT-T | mission, enemy, terrain, troops, and time available |
| MICLIC | mine-clearing line charge |
| MIJI | meaconing, intrusion, jamming, and interference |
| min | minute(s) |
| mm | millimeter(s) |
| MOPMS | Modular Pack Mine System |
| MOPP | mission-oriented protective posture |
| MP | military police |
| MRB | motorized rifle battalion |
| MRC | motorized rifle company |
| MRE | meal, ready-to-eat |
| MRP | motorized rifle platoon |
| MSR | main supply route |
| MST | maintenance support team |
| | |
| N | no |
| NA | not applicable |
| NAI | named area of interest |
| NBC | nuclear, biological, chemical |
| NCO | noncommissioned officer |
| NCS | net control station |
| No | number |
| NP | nonpersistent |
| | |
| OBSTINTEL | obstacle intelligence |
| OCOKA | observation and fields of fire, cover and concealment, obstacles, key terrain, and avenues of approach |

| | |
|---|---|
| OP | observation post |
| OPCON | operational control |
| OPLAN | operation plan |
| OPORD | operation order |
| OPSEC | operations security |
| ORP | objective rally point |
| oz | ounce(s) |
| | |
| P | persistent |
| PCC | precombat check |
| PCI | precombat inspection |
| PDM | pursuit-deterrent munition |
| PIR | priority intelligence requirements |
| PL | phase line |
| PLD | probable line of deployment |
| PLL | prescribed load list |
| plt | platoon |
| PMCS | preventive maintenance checks and services |
| POL | petroleum, oils, and lubricants |
| psn | position |
| | |
| R&S | reconnaissance and surveillance |
| RAAM | remote antiarmor mine |
| recon | reconnaissance |
| reqd | required |
| | |
| S&P | stake and platform |
| S1 | Adjutant (US Army) |

| | |
|---|---|
| S2 | Intelligence Officer (US Army) |
| S3 | Operations and Training Officer (US Army) |
| S4 | Supply Officer (US Army) |
| SCATMINE | scatterable mine |
| sec | second(s) |
| SEE | small emplacement excavator |
| SHELREP | shelling report |
| SITREP | situation report |
| SOI | signal operation instructions |
| SOP | standing operating procedure |
| SOSR | suppress, obscure, secure, reduce |
| SPOTREP | spot report |
| T/C/S | tasks, conditions, and standards |
| T72/80 | Soviet tank |
| TA | table of allowance |
| TC | tank commander |
| TCP | traffic-control post |
| TDP | turret-down position |
| TF | task force |
| TLP | troop-leading procedures |
| TOC | tactical operations center |
| TRP | target reference point |
| TTP | tactics, techniques, and procedures |
| UMCP | unit maintenance collection point |
| US | United States |

| | |
|---|---|
| USAF | United States Air Force |
| VX | VX |
| w/ | with |
| WO | warning order |
| WP | white phosphorous |
| wt | weight |
| XO | executive officer |
| Y | yes |

# Index

# References

## SOURCES USED

These are the sources quoted or paraphrased in this publication.


**Army Publications**

FM 3-4. *NBC Protection* (FMFM 11-9). 29 May 1992.

FM 3-5. *NBC Decontamination* (FMFM 11-10). 17 November 1993.

FM 5-34. *Engineer Field Data*. 14 September 1987.

FM 5-100. *Engineer Combat Operations*. 22 November 1988.

FM 5-103. *Survivability*. 10 June 1985.

FM 7-8. *Infantry Rifle Platoon and Squad*. 22 April 1992.

FM 20-32. *Mine/Countermine Operations*. 30 September 1992.

FM 71-100. *Division Operations*. 16 June 1990.

FM 90-7. *Combined Arms Obstacle Integration*. 29 September 1994.

FM 90-13-1. *Combined Arms Breaching Operations*. 28 February 1991.

FM 21-60. *Visual Signals*. 30 September 1987.


## DOCUMENTS NEEDED

These documents must be available to the intended users of this publication.


**Department of the Army Forms**

DA Form 1155. *Witness Statement on Individual*. 1 June 1966.

DA Form 1156. *Casualty Feeder Report*. 1 June 1966.

DA Form 1355. *Minefield Record*. March 1987.

DA Form 1355-1-R. *Hasty Protective Minefield Record (LRA).* July 1975.

DA Form 2028. *Recommended Changes to Publications and Blank Forms.* 1 February 1974.

DA Form 2404. *Equipment Inspection and Maintenance Worksheet.* 1 April 1979.

Made in the USA
Monee, IL
15 September 2026

63196022R00066